"At long last, a book that effectively unravels the mysteries of
natural plasters for all builders interested in creating their own beautiful,
non-toxic, earth-friendly homes. Cedar and Dan have compiled this hard-to-find,
practical information into a single source that speaks from their life's work
experience and focus on healthy building alternatives. This book will expand
the consciousness of everyone who likes to 'play in the mud'!"

— Steve Kemble, Co-producer of
"How To Build Your Elegant Home with Straw Bales"

The Natural Plaster Book

earth, lime and gypsum plasters for natural homes

The
Natural Plaster
Book

earth, lime and gypsum plasters for natural homes

CEDAR ROSE GUELBERTH & DAN CHIRAS

Foreword by Matts Myhrman & Judy Knox
Illustrations by Deanne Bedner

NEW SOCIETY PUBLISHERS

Cataloguing in Publication Data:

A catalog record for this publication is available from the National Library of Canada.

Illustrations by Deanne Bedner.

Cover design by Diane McIntosh. Plaster cat photograph © Kiko Denzer, sculptor, and author of *Build Your Own Earth Oven* (Hand Print Press). Building photo courtesy Catherine Wanek.

Printed in Canada by Transcontinental Printing.

New Society Publishers acknowledges the support of the Government of Canada through the Book Publishing Industry Development Program (BPIDP) for our publishing activities.

Paperback ISBN: 00-86571-449-5

Inquiries regarding requests to reprint all or part of *The Natural Plaster Book* should be addressed to New Society Publishers at the address below.

To order directly from the publishers, please add $4.50 shipping to the price of the first copy, and $1.00 for each additional copy (plus GST in Canada). Send check or money order to:

New Society Publishers

P.O. Box 189, Gabriola Island, BC V0R 1X0, Canada

1-800-567-6772

New Society Publishers' mission is to publish books that contribute in fundamental ways to building an ecologically sustainable and just society, and to do so with the least possible impact on the environment, in a manner that models this vision. We are committed to doing this not just through education, but through action. We are acting on our commitment to the world's remaining ancient forests by phasing out our paper supply from ancient forests worldwide. This book is one step towards ending global deforestation and climate change. It is printed on acid-free paper that is **100% old growth forest-free** (100% post-consumer recycled), processed chlorine free, and printed with vegetable based, low VOC inks. For further information, or to browse our full list of books and purchase securely, visit our website at: www.newsociety.com

NEW SOCIETY PUBLISHERS   www.newsociety.com

DEDICATION

To our children:

Cedar's daughter and son, April and Summer,

and Dan's two sons, Forrest and Skyler.

TABLE OF CONTENTS

Acknowledgments

MANY PEOPLE HAVE HELPED US throughout this project, answering questions over the phone or via e-mails, or through their written materials or interviews. Many thanks to all of them, including Matts Myhrman, Judy Knox, Steve Kemble, Carol Escott, Chris Magwood, Bob Campbell, Kaki Hunter, Doni Kiffmeyer, David Eisenberg, Bill and Athena Steen, Carole Crews, Shay Salomon, Catherine Wanek, Linda Smiley, Ianto Evans, Robert Laporte, Paula Baker-LaPorte, Kai Stapelfeldt, Charmaine Taylor, Niko Horster, Michael Smith, Kiko Denzer, Frank Andersen, Carmen Vasquez, Albert Andrews, Jr., Dorothy Andrews, and Reto Messmer. We'd like to thank Keith Lindauer, Johnny Weiss, Linda Smiley, Catherine Wanek, Doni Kiffmeyer, Kaki Hunter, Jean-Louis Bourgeois, and Edith and Alex Forrester for the photos they've provided. If we have forgotten anyone, we apologize profusely.

A world of thanks to our families for their love and support through the many months of writing and rewriting. And, of course, we'd be remiss if we didn't acknowledge the folks at New Society Publishers. At the helm of this splendid ship is Chris Plant, whose patience, guidance, and unwavering support have been nothing short of phenomenal, for which we are eternally grateful. Many thanks to Deanne Bednar for her delightfully skillful drawings which grace the pages of this book. We would like to acknowledge with tremendous gratitude the rest of the staff at New Society, especially Greg Green, production manager and art director, and Sue Custance, production coordinator, for their assistance in the preparation of this book.

Foreword

THERE IS AN OLD CHINESE SAYING, "If we don't change direction, we will surely end up where we are headed." The world we inhabit and are irreversibly interconnected with is in deep trouble; every natural system on earth is in decline, some precipitously. We, as human beings, must choose to claim responsibility for the cumulative results of the choices that have brought us to this critical point in the history of our planet. It will take enormous effort and courage to move beyond our ordinary way of doing things; it will take each and every one of us, acting on choices that arise from our deepest inclinations to affirm life. What's at the heart of our human existence is that there is an essential part of us that yearns — even clamors — to champion the breakthroughs necessary to restore and sustain life.

There is no area of life needing dramatic change more than the way we in the "developed" world go about housing ourselves — housing that is often toxic to both the planet and its inhabitants. In the United States, our buildings account for 40% of all material and energy use, 35% of greenhouse gas production, and 28% of municipal solid waste. Since the 1940s, floor space per person in new homes has nearly tripled. Our houses demonstrate many unhealthy habits: use of energy-consumptive, unhealthy, man-made materials; ecologically destructive misuse of natural materials; decades of mortgage loan debt, the payment of which requires excessive amounts of time and energy — energy needed for ourselves, our families, our communities; the family's almost complete disconnection from the design/building process of their homeplace; and, overarching all, our seemingly insatiable need for way more than enough to meet our basic needs.

As leaders of the straw-bale construction revival, we constantly asked ourselves how we could best demonstrate and inspire a move from egocentric to ecocentric buildings. As champions of natural building, the transformative power of our work together resides in developing our ability to inform and inspire others — to build our technologies and workplaces into bridges of learning and demonstration for the legions of people who can't imagine how we can get from here (serving the imperatives of a consumer-driven, growth-oriented, anthropocentric world) to there (creating just, sustainable societies that bring the human species into balance with itself and the planet). Real champions dedicate themselves to reflecting hope out into a distressed world — not blind hope, but hope that rises out of developing and teaching real and do-able ways to meet our basic human needs within a restorative and sustainable framework.

Our hats are off to champions Dan Chiras and Cedar Rose Guelberth, who have spent years gathering information, learning, teaching, trying out new ways of doing things, discarding what doesn't work, improving those things that do work, collaborating with others in the natural building community, and, most impressively, making the effort to turn their knowledge into usable and available tools for all of us.

The Natural Plaster Book is an open invitation to the champion in all of us; to add its hard-earned information to our tool kits, try it out in our homes and communities, modify, add, detract, collaborate with others, teach, learn, share — to join in the step-by-step, conscious, choice-filled, joyful, and hopeful journey from here to there.

Judy Knox and Matts Myhrman
Out On Bale
Tucson, Arizona

Introduction

CEDAR AND I MET IN JUNE OF 1999 at a workshop on natural building in the mountains of Colorado in a largely defunct mining town known as Rico. She was teaching, as usual, and I was attending workshops and lectures, hungrily gathering information for a new book, *The Natural House*, slated for publication the following year. The sun shone bright, unabashed. The air was warm and inviting, and the Colorado sky was a flawless blue dome above our heads, free of the haze and pollution you see most everywhere else in this country. I was heading into town to hear Cedar speak at the local theater which was our meeting place. She asked for a ride, and off we went.

In the mile or two from our campground to the funky theater where the slide shows and lectures were given, we talked up a storm. In that short distance, it became apparent that she was a rather special person, full of enthusiasm, kindness, and an expanse of knowledge on natural plasters that could fill a book or two. Before we had parked in town, a couple of miles from the site where we all camped and engaged in the hands-on portion of the three-day workshop, we'd broached the subject of working on a natural plasters book together. I liked the idea, as I'm always on the lookout for new and exciting topics to research and write about, but I must say I also felt some trepidation. To say I didn't know much about plasters at the time would be the understatement of the year. I had worked with unnatural wall finishes — cement and synthetic stucco — and had read a little bit about earthen plasters, but at the time I would have had trouble writing a coherent paragraph on earthen plasters, let alone lime and gypsum plasters.

With some trepidation, I suggested "Maybe we should work on a book together." When we left the workshop, the idea still swirled in my head. A year later, when we had time to hammer out the details, we embarked on this project. You're holding proof that we managed to forge a fruitful partnership that melded Cedar's vast knowledge on the subject with my modest skills at research and writing.

This book is a labor of love — the first comprehensive book on natural plasters for natural buildings. It took much longer to write than either of us had ever imagined, but the process — even though grueling at times — was ultimately successful. I am thankful for the opportunity to translate Cedar's gained knowledge — along with knowledge of a great many others who communicated with us through written and spoken word — into a book that offers insight, guidance, and enthusiasm for a subject that ranks among the top that I've had the privilege to tackle since the 1980s when I left my full-time university position to pursue a life of independent research and writing.

As you will soon see, this book describes natural plasters on natural buildings. Although much of our attention focuses on earthen plasters on straw bale homes, there is a great deal of information on lime and gypsum plaster, and on making and applying plasters to a wide range of natural buildings. Our goal throughout the book is to provide you with a firm conceptual understanding of natural plasters, one that allows you to tackle virtually any project with confidence, and to give you important details that will make any plaster job more rewarding, safer, and more successful.

How This Book is Organized

We begin our book with an overview of natural building. Chapter 1 is designed to help the reader understand the various building systems we will refer to in the book. This discussion is followed by a useful overview of natural plasters in Chapter 2, which provides a little background information on plasters that's essential to your understanding of traditional plasters. In Chapter 3, we discuss important details of the planning, design, and construction of natural homes, especially straw bale homes — details required for a successful plaster job. Be sure to read this chapter: it is vital to your success.

In Chapter 4, we begin our in-depth look at earthen plasters. We'll explore each of its components — sand, clay, silt, and fiber — and the role each plays in an earthen plaster. You will learn how to test soil — usually subsoil — to see if it is appropriate for making earthen plaster or how it needs to be altered. Next, we examine plaster additives — substances you can add to an earthen plaster to make it easier to work with and more durable and water-resistant.

In Chapter 5, we turn our attention to site preparation and mixing plasters. We'll provide guidelines for making your job site clean, efficient, and safe; explain how to prepare materials and mix plasters; and provide an overview of the function of the various plaster coats.

In Chapter 6, we look at the application of earthen plasters on straw bale homes, starting with the prep coats, then proceeding to the layers of plaster itself. You'll learn more about mixing plasters and the techniques used to apply each coat. As in other chapters, we'll describe the tools you will need.

Next, in Chapter 7, we will focus our attention on wall finishes. You will see how you can add color to earthen plaster walls via alises, litema, clay finish coats, and natural paints.

In Chapter 8, we'll explore the world of lime plasters and in Chapter 9, gypsum plasters. Then, in Chapter 10, we will discuss what you need to know and do to successfully plaster walls made of cob, adobe, rammed earth, straw-clay, rammed earth tire, and earthbags. Even though only one of these methods of construction may be the type of building you are interested in, we urge you to read this book from cover to cover. Much of what you learn early on, while focused primarily on straw bale homes, does carry over to other natural homes.

Finally, in the Resource Guide at the end of the book we provide a comprehensive listing of publications (books, articles, newsletters), videos, organizations, suppliers, and workshops.

We welcome newcomers to the natural building movement and hope this book helps in many ways, providing a solid conceptual background in addition to details on processes and materials that will help you to become a successful natural plasterer — or will improve the knowledge and skills of those of you who have already begun to dabble in this wonderful craft.

In closing, we would like to point out that slopping around in the mud may not seem like the most civilized thing a human could do. However, if that mud is an earthen plaster destined to adorn the walls of a natural home, this pursuit may just turn out to be one of the most enlightened acts of civility you can engage in...that is, if you care about the future of our planet, our children, and the many species that share this planet with us. You will see why shortly.

Dan Chiras
Evergreen, Colorado

Cedar Rose Guelberth
Carbondale, Colorado

CHAPTER 1

Welcome to the World of Natural Building

F OR VIRTUALLY ALL OF HUMAN HISTORY, our ancestors have lived in shelter fashioned from
locally available materials. Earthen materials were one of the most popular. Even
today, approximately half of the world's people inhabit shelters fashioned from clay-rich
dirt harvested from the Earth's crust. Such shelter not only protects people from the elements, it can provide extraordinary comfort, even in rather harsh climates. Earthen
building materials also create a close connection to the Earth with calming and healing
effects.

Today, however, an increasing number of
homes are being built from synthetic or highly
processed natural materials. Many modern materials release toxic substances into our homes,
inadvertently poisoning the very people these
homes were designed to protect. Their harvest and
production also damage the environment.

Most contemporary shelter is also less than
optimal for maximum human comfort. For one,
most new homes tend to create a sterile, straight-
edge environment — so unlike the natural world
and so devoid of soul. Placed in cities and suburbs,
our homes also tend to isolate us from nature. The
closest most people get to nature anymore is a romp on the pesticide-sprayed lawn or
in a local park with the family dog. Coming from generations of people who were con-
nected to nature, many of us respond poorly to the isolation from our environment.

There is a far more healthful way to create shelter. It is called natural building. With
thick, protective walls fashioned from earth and fiber, natural homes typically offer soft
lines and delightfully curved, even sensuous, walls. Finished with a sumptuous earthen

ALBERT M. ANDREWS, JR.

*1-1: Locally available materials
have been used throughout
the world for virtually all of
human history to build homes
and other structures like these
stone and earthen buildings
in old Jerusalem.*

1

plaster, these homes often evoke feelings of security, harmony, and peace. Built with features likely to be found in the natural environment, these homes help to connect us to the Earth, the source of life.

Natural homes provide a nurturing and supportive environment for people. They are a kind of mental salve to battered senses in a high-stress world. Natural homes provide us an opportunity to relax and rejuvenate after a hard day's work. For those who work in a natural home or in a natural office building, immersion in this setting provides a daily nurturing environment — far more productive than most contemporary buildings. As many readers know, prolonged stress can impair the human immune system and endanger our health. By reducing stress, natural homes may help our immune systems function at their peak and thus protect our health.

Natural building is good for our bodies and our minds, but the list of benefits does not end here: it also pays huge environmental dividends. Most are made from locally available materials transported to the building site using far less fossil fuel energy than those needed to build a modern stick-frame structure. Less energy means less pollution. Although locally harvested materials can create small isolated pockets of damage, these can be repaired quite easily. The small hole dug to extract clay for an earthen plaster, for instance, may be converted to a frog pond, or filled in with topsoil and replanted, leaving no evidence of earlier disturbance. By building with locally available natural materials, you can reduce the use of highly manufactured materials whose production often causes extraordinary environmental damage. The list of benefits goes on, but the point should be clearly evident: by building a home that nourishes body and mind, we protect and replenish our environment. Planet care, we must not forget, is the ultimate form of self-care.

1-2: The earthen plastered walls of this straw bale home are inviting and soothing to the soul, in part because they connect us to the Earth.

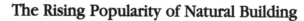

CEDAR ROSE GUELBERTH

The Rising Popularity of Natural Building

Although natural home building has declined sharply in modern times, it is making a strong comeback. Straw bale building has been a pivotal player in this resurgence. The revival of straw bale building, in turn, can be traced in large part to an article written by Roger Welsch which was published in a fairly obscure publication, *Shelter Magazine*, in the 1970s. Penned primarily for the back-to-the-land folks, this article provided an overview of indigenous and off-beat ways of constructing shelter. Although few would have predicted it, the article also inspired a dramatic resurgence in straw bale construction, a technique that originated nearly 100 years earlier in the wind-swept grasslands of the Sand Hills of western Nebraska. The resurgence began in the early 1980s, slowly at first. Then, in the 1990s, it gained a momentum which continues today in North America, Europe, Australia, Russia, Japan — all over the world!

Straw bale building offers numerous benefits and attracts a wide spectrum of people. Most people are attracted to straw bale building's promise of reduced energy bills. Energy efficiency, combined with passive solar heating and passive cooling, can result in dwellings that use little, if any, outside energy — and thus contribute mightily to cleaner air. In most parts of the world, straw is locally available. Far less energy is required to ship a truck load of straw from a farm 20 miles (12 kilometers) away than to ship wood from distant forests 1,000 - 2,000 miles (600 – 1200 km) away. This, too, adds to its appeal.

Straw bale construction helps reduce air pollution in other ways as well. In many parts of the world, after grains are harvested, the straw is burned off to make cultivation easier and to return nutrients to the soil. So, advocates point out, building houses out of straw is a great way to put an agricultural waste product to good use and to reduce air pollution.

Straw bale construction also provides a way of reducing wood use. For those interested in helping to protect our vanishing forests, this is a major plus. In addition, straw bale replaces potentially toxic or earth-unfriendly insulation materials — for example, fiberglass containing formaldehyde or foam made from ozone-depleting chemicals. It is, therefore, a more healthful way of creating shelter.

Straw bale building is appealing to people who want to construct their own homes. That's because building walls from straw bales is relatively easy.

Straw bale construction can also reduce the cost of building a home, although this is not always the case. If designed and built well, straw bale homes save money in energy bills and thus dramatically reduce living expenses.

Over time, we've discovered another benefit of straw bale construction: straw bale homes can be exceedingly beautiful. The adjectives elegant, graceful, and breathtaking describe many well-crafted straw bale homes.

1-3: The Martin/Monhart home, built in Arthur, Nebraska in 1925, demonstrates the elegant simplicity and endurance of straw bale construction.

THE ECONOMICS OF STRAW BALE CONSTRUCTION

Straw bale homes range in price, depending on how much work an owner does and how much he or she must contract out. Cost also depends on the complexity of the design. The more complex, the more costly the home. In addition, cost depends on details — for example, the type of tile and cabinetry and the amount of finish work. Straw bale homes can range from as little as $50 to $150 per square foot.

Straw bale building can be fun, too, and is often carried out communally with wall-raising parties or workshops that build personal relationships as well as shelter.

Despite what critics say, straw bale walls, if designed and built well, rarely have problems and are extremely durable. Straw bale walls are also pest- and fire-resistant, because finished walls are coated with a thick layer of plaster that prevents pests and fire from reaching the bales. Even if fire penetrates the plaster, straw bales are tightly compacted and burn poorly due to a lack of oxygen.

With thick walls of straw, these homes are quiet. In addition, straw bale building is an approved form of construction in many jurisdictions. Mortgages and insurance are also readily available in many locales. Resale values appear to be quite good, too. If the energy supplies dwindle, the value of a highly energy-efficient straw bale home could easily exceed a comparably sized structure made from conventional materials.

Despite its many benefits, straw bale building does have a few shortcomings — but then so does every other form of construction. Straw is the shaft of cereal crops, such as wheat. Many farmers who grow wheat and other cereal grains use lots of irrigation water and pesticides. (Use pesticide-free bales if possible.) Using straw, rather than plowing it under, reduces the return of nutrients to the soil. (Buying bales from farmers who manage their farms well can lessen this impact.) Straw bale building is not always as inexpensive as some would have you believe. (Walls are only a small portion of the total cost of building a home; many other factors determine the total cost.) If not designed well (especially with regard to protection from water), straw bale walls can mold and deteriorate, as in any construction system.

1-4: Straw bale raisings like this one promote community, reminiscent of the barn raisings of days past. They also help to teach building skills to others.

DAN CHIRAS

In closing, straw bale construction provides many benefits. While there are some things to be aware of, straw bale construction offers a wonderful way to build sustainable homes and has a bright future.

Natural Building: Creating Homes from Earth and Fiber

Although many readers may be familiar with straw bale, numerous other natural building materials are gaining in popularity in recent years. In fact, there are currently over a dozen different natural or alternative building systems, among them rammed earth, adobe, straw-clay, earthbags, and cob. Although there are many earth-friendly building alternatives, walls are generally built from two materials, earth and fiber (for example, straw). Stone and wood are often used as structural components (foundations in the case of stone, framework in the case the wood). In many homes, natural builders are using a combination of natural materials. For instance, they may build exterior walls out of straw bales and interior mass walls for passive solar heating from adobe blocks, rammed earth, or cob. Interior divider walls may be built from straw-clay. (We'll describe each of these options shortly.) Many natural homes are finished with natural plasters, too.

No matter what natural building material is used, they all share two common features: they're produced by natural processes and they're locally available. Because they're made by natural processes, they're renewable. Straw and soil, for instance, are both renewable resources — although soil formation takes a long time.

CEDAR ROSE GUELBERTH

Using such materials allows us to build homes that tread lightly on the Earth. Moreover, many natural building materials are conducive to highly energy-efficient design. They're also ideal for passive solar heating and cooling. Natural building therefore is gentle on the environment during construction and during the life span of the house. Once its useful life is over, the materials used in building a natural home can easily revert back to their former state.

Because we'll be talking about natural plasters for earthen homes as well as straw bale structures, we begin with a survey of the major natural building systems to which plasters are applied.

1-5: One of the great advantages of building with natural materials, including natural plasters, is that when a house reaches the end of its useful life, the materials can easily revert back to their original state, as seen in this photo of disintegrating adobe bricks. These materials can even be recycled into new buildings.

If you want to learn more about a particular building system, we refer you to Dan's book, *The Natural House: A Complete Guide to Healthy, Energy-Efficient, Environmental Homes*, published by Chelsea Green. This book describes each natural building technique in detail and contains an extensive Resource Guide that is updated and expanded on Dan's web site: www.chelseagreen.com/Chiras. (Be sure to capitalize the C in Dan's last name or it won't let you in.) The Resource Guide at the end of this book also lists numerous publications and videos on natural building that readers will find useful.

Earthen Homes

Earthen homes include structures made from adobe, cob, rammed earth, rammed earth tire, and earthbags. We'll take a brief look at each one.

ADOBE. One of the most widely recognized natural building materials is adobe. Adobe has been used for thousands of years, and is still being used throughout the world: in China, the Middle East, Northern Africa, South America, Central America, and the United States. Today, tens of thousands of adobe structures still stand, providing comfortable shelter. Many new structures are being built each year.

Adobe bricks are made from a mix of clay-dirt, sand, and straw. Wetted, mixed, and then poured into forms to create bricks that are dried in the sun, adobe is an ideal building material — forgiving and easy to learn, and is suitable for many climates.

1-6: Adobe can be used to build a variety of different architectural styles, not just the typical style seen in the U.S. Southwest which is often associated with adobe.

JOHNNY WEISS - SOLAR ENERGY INTERNATIONAL

Discard any preconceived notions you might have about adobe being restricted to desert climates. Although adobe does indeed perform well in hot, dry climates with cold nights, adobe homes are found in many other regions as well, including some rather chilly climates. In New York State alone, for instance, researchers have documented at least 40 adobe homes in a nine-county region, covering about half of the state. These homes were built well over 100 years ago. Even Paul Revere's home in Boston was built from adobe bricks. In cold climates, however, steps must be taken to insulate adobe walls to prevent heat loss.

Besides being adaptable to a variety of climates, adobe can be used to build homes of many different architectural styles. In the United States, adobe homes are commonly built in the southwestern style. Worldwide, you'll see adobe homes built in a variety of architectural designs.

Adobe provides a secure and stable wall system. Adobe bricks are laid on a foundation in a running bond (an overlapping pattern which provides strength). They are then mortared in place using an earthen mix similar to that used to make the bricks and usually coated with a protective layer of earthen plaster, which expands and contracts at the same rate as adobe bricks. Earthen plasters not only protect the adobe bricks, they add a measure of beauty.

COB. Another natural building technique is English cob, which, like adobe, consists of a mix of clay-rich dirt, sand, and straw. But rather than using the mix to create blocks, the material is applied, often in loaf-shapes, to the foundation and wall directly by hand or by the shovelful. (Cob is the English word for a lump or rounded mass.)

In cob construction, walls are molded and shaped by hand. As a result, cob lends itself to sensuous curved walls, arches, and niches. Cob homes can be as much an expression of one's artistry as a place to live.

1-7: Bales and adobe blocks are laid in a running bond pattern to ensure the stability and strength of walls.

Exterior cob walls are usually 24 inches thick and quite durable. (Interior walls are usually thinner.) When cob dries, it becomes hard and strong like sandstone. Cob walls are often lime washed, lime plastered, or coated with an earthen plaster for added protection and beauty.

Cob is ideal for owner-builders and is easy enough for children to master. Most of the work is done by hand or with simple hand tools. Building with mud is fun. Like adobe, cob works in many climates, so long as precautions are made to protect walls from the elements, especially driving rains. (Large overhangs and porches work well.)

1-8: These beautiful cob homes in Great Britain have, like thousands of similar buildings, been continuously occupied for hundreds of years.

1-9: Cob and other forms of natural building promote artistry and freedom of expression not possible with conventional building materials.

Like adobe, cob is a time-tested building technique. It has, for instance, been used extensively in southern England, where tens of thousands of cob homes remain standing, and are still occupied after 500 to 700 years!

As with many other natural building techniques, workshops, books, and videos on cob building are available. Cob is often used in conjunction with other building techniques, such as straw bale. In such instances, cob serves as an excellent filler. Workers often apply cob in nooks and crannies or to sculpt interior features such as benches or niches. Cob can also be used to build interior mass walls required for passive solar heating and cooling.

1-10: Worker using a pneumatic tamping device compacts slightly moistened earth in a form mounted on a foundation to create a solid structural rammed earth wall.

1-11: (right) Rammed earth walls are thick and massive; they work well in arid climates to buffer against hot summer temperatures.

JOHNNY WEISS - SOLAR ENERGY INTERNATIONAL

JOHNNY WEISS - SOLAR ENERGY INTERNATIONAL

RAMMED EARTH. Another traditional building technology is rammed earth. As its name implies, this is made from earth that's "rammed" or packed between forms. Workers begin by erecting wooden or steel forms on a hefty foundation. A little moistened, clay-rich subsoil is shoveled into the forms, then tamped — either by hand or with a pneumatic tamper. Additional subsoil is added, then tamped, and so on until a wall is formed.

Soon after the last bit of tamping has been completed, the forms are removed. Exterior rammed earth walls are usually 12 to 24 inches thick. Interior walls can also be constructed from rammed earth, but to save floor space they're usually narrower.

Rammed earth is an ancient technique. In China, rammed earth buildings have been discovered which date back to the 7th century BC. Parts of the Great Wall of China, begun over 5,000 years ago, were also made from rammed earth. Ancient rammed earth buildings are found in North Africa and the Middle East, where the practice continues today. Many rammed earth structures can also be found in France, where rammed earth was the dominant form of building 2,000 years ago. Historians believe the Romans introduced this building technology to the picturesque Rhone River Valley.

Rammed earth building is taking off in the United States, particularly in California, Arizona, and New Mexico. It is growing in popularity in New Zealand, too, but nowhere is the technique more popular than in parts of Western Australia where a quarter of all new homes are built from rammed earth.

Like other natural building techniques, rammed earth homes provide massive exterior walls, which perform ideally in hot, dry climates. In colder climates, rammed earth serves as thermal mass. As long as there is a good supply of heat, either from the sun or some other source such as a masonry heater, the walls will stay warm. If not, they may require insulation to prevent excess heat loss. Earth is not a great insulator in itself.

Finished rammed earth walls can be breathtaking to behold. Some builders even use pigments in layers to create a stratified look. These walls are often left unplastered — it would be a shame to hide such beauty. However, like adobe and cob, rammed earth walls are sometimes plastered — at least externally — to provide additional protection or to fit better into the neighborhood.

1-12: The Earthship is built from used automobile tires rammed with earth, then coated in plaster. This home was designed and built by Peter Kolshorn in Taos, New Mexico.

EARTHSHIPS AND TIRE HOMES. While we're on the subject of rammed earth, we'd be remiss if we didn't mention Earthships and other tire homes, an alternative building technique using recycled materials. Earthships are essentially rammed earth homes. However, instead of using rigid flat forms to build walls, builders use automobile tires, which are laid on compacted subsoil or a foundation. Each tire is filled with dirt then compacted with steady blows of a sledge hammer or the powerful packing action of a pneumatic tamping device. In a short while, the tire bulges, full of tightly compacted dirt. (A 15-inch tire will hold up to 300 or 350 pounds of dirt!) The next tire is then packed, and so on down the line. After the first row is completed, a second row is placed on top in a running bond (overlapping pattern), and these tires are then filled with dirt and compacted. Six to eight rows of tires make up a wall. After the wall is completed, it can be plastered, often with mud or cement stucco.

1-13: Earthbags are filled with a slightly moist soil, as shown here, then tamped in place to compact the soil, producing hard, brick-like structures ideal for building walls.

Earthships are the creation of builder Michael Reynolds from Taos, New Mexico. They're designed to be self-contained vessels that sail us into the future with little environmental impact, hence the name Earthships. Tires can also be used to make more conventional-looking homes, which have the amenities of an Earthship, yet fit into an established neighborhood more readily.

Although not everyone will want to live in an Earthship, readers are encouraged to study Reynold's ideas on integrated design, self-sufficiency, and sustainability. His books are listed in the Resource Guide.

EARTHBAGS. While we're still on the subject of ramming earth, there's another natural building technique worth mentioning. It's called earthbag construction.

1-14: Earthbags placed over forms create wonderful arches for windows, as shown here.

1-15: Good foundations and roofs protect earthen and straw bale walls from moisture, which is essential to the longevity of both walls and plasters.

One of the newest natural building methods, Earthbag construction is both versatile and durable. Polypropylene bags (used for sand bags or the kind bulk rice comes in) are typically used, although other types (such as burlap and jute) are also suitable. The bags are filled with a moistened soil mix containing sand and clay-rich soil. The bags are laid down on a foundation, and then tamped. As the builder pounds the bags, the dirt compacts. When dry, it becomes hard as a rock. Earthbags are laid in a running bond, then covered with mud plaster which sticks surprisingly well to the surface. Lime plaster can also be used on Earthbag walls.

Earthbags are ideal for making round structures with dome roofs or creating vaults. Earthbags are also used to build foundations for other types of natural homes, although special precautions need to be made to prevent moisture from seeping up through the bags. Earthbags can, for instance, be stabilized with a little cement (five percent) to increase moisture resistance of foundations, although an initial course or two of bags containing crushed rock work even better to reduce water migration up into overlying bags. Books and other materials on Earthbag construction are listed in the Resource Guide.

Straw Walls

The buildings we've discussed so far rely primarily on earth to fashion exterior walls. Although cob and adobe incorporate some straw in the walls, the walls are still primarily earthen and have a low insulating value. Far more insulation is provided as the fiber content of the walls increases. One of the most fiber-intensive is the straw bale home.

STRAW BALE. Straw bale homes are made from straw — not hay — bales laid in a running bond pattern. To prevent the bales from getting wet, they must be placed on a well-designed foundation and protected by ample roof overhang and exterior plaster or some kind of siding. These and other measures are described more fully in Chapter 3.

To enhance lateral stability, builders may drive long pins (rebar or bamboo) through the center of the bales, three rows at time. Most builders, however, now pin straw bale walls externally — that is, they attach bamboo or some other pin along the outside of the walls. External pins run from the top plate to the bottom plate and thus help make a straw bale wall more rigid.

Contrary to what many outsiders think, straw bale walls are strong — so strong, in fact, that they can support the weight of roof with only a top plate in place to distribute the load. Straw bale walls built to support the roof directly are called load-bearing walls.

In many cases, builders use straw bales merely as an infill. A wooden frame is built to support the roof. The straw bales are inserted between vertical members of a wooden frame — typically either a timber frame or a post-and-beam structure. Straw bales can also be "wrapped around" a wooden frame. Such walls are called nonload-bearing straw bale wall — because the frame bears the roof load, not the bales.

Once the walls are up, they're usually plastered. Cement stucco, earthen plaster, lime, and gypsum have all been used. For reasons we will make clear in subsequent chapters (Chapters 2, 3, and 10), we recommend against the use of cement stucco on straw bale — or any other natural building.

Straw bale construction is becoming so popular that there are numerous books, videos, and workshops on the subject, many of which talk a little about plastering, too. We highly recommend you explore these resources.

1-16: Straw-clay homes like this one in Germany have been around for hundreds of years.

STRAW-CLAY. Another straw-intense building material is straw-clay, sometimes referred to as light straw-clay. Straw-clay has a long and rather successful history of use in Europe. In fact, throughout Europe you'll find 500- to 1,200-year-old straw-clay homes that are still occupied.

Straw-clay is made from straw mixed with a clay slip — a soupy mix of finely screened clay-rich dirt and water. To make straw-clay, one simply throws some clay slip onto straw and mixes it a bit, so the clay lightly coats the strands of straw. When fully mixed, the straw-clay is placed in forms as in rammed earth, then tamped to compact it. Forms are usually only about 18 – 24 inches (45 – 60 cm) high. Once a section has been completed, a new form is added and then filled and packed. When it is completed, the first form is removed, and leapfrogged — that is, moved above the second form.

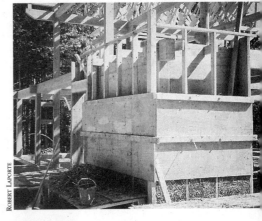

Straw-clay is used in conjunction with some kind of structural framing, for example, timber framing or stick-frame construction. Packed straw clay is nonstructural, meaning it does not support weight. Thus, the straw- clay is a form of insulation packed between structural members of the wall. When the forms are filled, they are removed, and the wall then dries. After the walls are dried, they can be coated with plaster. Earthen plasters work extremely well on straw-clay walls.

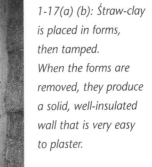

1-17(a) (b): Straw-clay is placed in forms, then tamped. When the forms are removed, they produce a solid, well-insulated wall that is very easy to plaster.

1-18: Rarely used in North America but still widely practiced in many parts of the world, wattle-and-daub consists of bamboo, willow, or wood lath woven together to form a matrix between framing members. A cob-like mix of clay dirt, sand, and straw is then applied to the surface.

Straw-clay is commonly used to construct thick exterior walls, but may also be used to build narrower interior walls. Straw-clay is gaining popularity in North America and is being used in conjunction with other natural building methods. Some builders also use a mix of clay slip and wood chips in place of clay slip and straw for walls.

WATTLE-AND-DAUB. Wattle-and-daub is a very old building technique which uses bamboo or sticks that are woven together to form a matrix between framing members. Though rare in North America, wattle-and-daub is still widely practiced in many parts of the world. In fact, there are as many ways of building with wattle-and-daub as there are tribal cultures around the world.

Wattle-and-daub systems fall into one of two categories. The first is a woven willow wall system in which willow branches (or branches from other trees) are woven to create a lattice between framing members. Mudders pack a wet, sticky cob mix over the willow lattice and build it out to a desired thickness. It then dries to form a smooth, cob-like exterior over which plaster is applied. The second system consists of two layers of woven materials, separated by a space of varying thickness. The wall cavity, between the layers of woven material, is packed with straw-clay, clay-slip and wood chips, wool, or other materials to create an insulated wall structure. Plaster is applied to the woven lath surface.

A Primer on Plaster

Now that you have an understanding of some of the most commonly plastered natural forms of building, let's take a look at plaster itself. What is it? What types of plaster are there? And what are its components?

What is Plaster?

Plaster is material used to cover walls — often both interior and exterior walls. It provides protection as well as texture and color. As you will soon see, there are many different types of plaster, each with its own pros and cons, and there are many ways of making plaster, which is applied in coats, each one with a different function.

Types of Plaster

Natural plasters — those made from natural materials that generally require very little processing — include three basic types: earthen, lime, and gypsum. Earthen plasters (the least processed of all natural plasters) are also referred to as mud, adobe, or (in Europe) loam plasters. As you shall soon see, earthen plasters can be applied on both interior and exterior walls, and are durable and weather-resistant, if done right.

Lime plaster, typically made of lime and sand, can also be used in interior and exterior applications — although it is typically applied on exterior walls where it provides excellent protection against the elements. Gypsum plaster, on the other hand, is a more delicate wall plaster. It is rather soft and water soluble, and is therefore typically reserved for interior work.

Outside of the realm of natural plasters, there are two other types of plaster worth noting: cement stucco and synthetic stucco. We'll talk a little about them in upcoming chapters. For now, suffice it to say that they're not advisable for natural homes. The primary reasons for avoiding these materials are that cement stucco wicks moisture into walls and it tends to crack, which allows moisture to enter. Once water gets into a wall, cement stucco tends to prevent it from escaping. Moisture causes straw to mold and rot and earthen materials to crumble. Synthetic stucco also traps moisture. In Chapter 4, we'll describe more of the hazards associated with the use of these materials in natural homes.

> ## STUCCO VS. PLASTER: WHAT'S IN A NAME?
>
> In North America, the term plaster refers to a wall covering made from earth, lime, or gypsum. Stucco is used to refer to a cement-sand or synthetic blend. In England, the term render is used instead of plaster.

Components of Plaster

All plasters and stuccos have several common features: they all contain a structural component, a binding agent, and some form of fiber. The structural component of plasters, except most gypsum plasters, is sand. Sand provides most of the volume of a plaster, and gives a plaster its structure. The binding agent bonds the sand particles to one another while the plaster is wet and after it dries or cures. As you will see, it is the type of binding agent that's used in a plaster or stucco that generally gives the material its name. In earthen plaster, it's the earthen clay that performs this function. In dung plaster, which is discussed in Chapter 7, manure is the binder. In lime plasters, lime serves as a binding agent. In gypsum plasters, gypsum is the binder. In cement stuccos, cement is the binding agent and in synthetic stuccos, artificially produced latex serves as the binding agent.

When the binding agents are mixed with water, they become adhesive. This, in turn, causes the sand grains to adhere to one another. The proper mix of binding agent and sand yields a malleable (pliable) plaster or stucco. The stickiness of the plaster allows the material to adhere to a wall. When the water evaporates, plasters and stuccos harden. The binding agents continue to hold the components together, creating a protective finish often capable of withstanding the forces of weather. Fibers are added to plaster to increase their strength and to provide reinforcement. Straw, cattails, wool, and wood fiber are commonly used in natural plasters.

Although they contain the same general components, plasters and stuccos vary considerably, giving each one its own unique properties, with advantages and disadvantages. We'll describe each of them, so you can develop a good understanding of your options and make wise decisions about which plaster to apply for each situation.

Quick Recap

People have been living in natural homes since the dawn of civilization. Although natural building has fallen by the wayside in many industrial nations, the bulk of the world's people still live in buildings made from earth or straw and many of these homes are hundreds of years old. Fortunately, there is a resurgence of the ancient methods and materials, which rely on locally available materials generated by natural processes.

Natural building results in shelter that is good for people in a wide variety of ways — for example, it is healthier and more nurturing than modern materials. It is also gentle on the Earth — not only during construction, but for the lifetime of the house, as many natural homes are amenable to passive solar heating and cooling and can last for centuries with little maintenance.

Although one of the most popular of all natural buildings is the straw bale home, many builders are creating homes from earthen materials, such as adobe, cob, rammed earth, rammed earth tires, and earthbags. Straw-clay and wattle-and-daub are used as well. Readers will find a growing body of information on these materials, as well as videos and workshops, to help them design and build the home of their dreams.

To protect and beautify the natural walls of their homes, many builders apply plasters inside and out. Natural plasters, including earthen, lime, and gypsum, are applied in coats. Plasters contain several common components: a binding agent and some form of fiber. Most plasters also contain sand.

Despite these common factors, plasters differ remarkably. Each has its own unique properties — and offers a set of advantages and disadvantages. Understanding each of them will help you make wise decisions when planning and building a house.

CHAPTER 2

An Introduction to Earthen Plasters

A S POINTED OUT IN CHAPTER 1, there are numerous types of plaster and stucco in use today. The oldest type of wall finish is earthen. It was used by many early builders to protect the walls of buildings made from natural materials. Today, many centuries-old buildings throughout the world still stand as a solemn testimony to the durability of natural building materials and their protective coatings of earthen plaster.

Earthen plasters were used by early builders to protect and beautify walls in large part because their components — sand, clay, and fiber — were readily available in many parts of the world. Variations in local soils led builders to seek the most effective materials and combinations of materials. Over time, however, builders began to experiment with various materials to create effective, durable plasters. One locally available substance was gypsum — a soft, natural sedimentary rock found throughout the world. It was used for a variety of purposes, among them interior plaster. Gypsum was also used to make elaborate decorations in walls and ceilings in Islamic and European buildings that have survived in good shape for thousands of years. Researchers believe that special additives may be responsible for their marble-like durability.

Builders discovered yet another locally available material — limestone. When crushed and heated in special kilns at very high temperatures, limestone is transformed into lump lime, a soft powdery substance. Water is added to lump lime, converting it to lime putty, and sand is added to produce lime plaster. When hydrated and mixed with sand, it makes an excellent plaster — as well as a mortar.

CEDAR ROSE GUELBERTH

2-1: Earthen plasters have a long history of use, protecting an assortment of natural buildings that have housed our ancestors for millennia.

15

2-2: In England, Portugal, Greece, and throughout the rest of Europe, lime plaster rose to a place of dominance, often replacing traditional earthen plasters. This Netherlands home demonstrates its durability as an exterior finish in harsh climates.

2-3: Ancient Roman baths built from stone and a lime-based mortar remain standing today, demonstrating the enduring strength of lime.

2-4: Cement stucco over earthen and straw bale walls can result in serious deterioration of the underlying wall, as evidenced here in this adobe building.

Lime plaster proved to be a durable finish material, useful for interior and exterior use. Lime has been used the world over for thousands of years. As Jocasta Innes points out in his book, *Applied Artistry: A Complete Guide to Decorative Finishes for Your Home,* "That so much of the ancient world survives is due to favorable climatic conditions and skilled construction, but also, in large measure, to the properties of building limes."

The Romans used lime to produce a concrete consisting of a mix of sand or small rocks and lime, hardened to produce a kind of artificial stone. They used this material alone to build vaults or as a mortar when constructing elaborate stone structures where the lime concrete often served as filling material between two surfaces of finished stone.

In England in the 1800s, lime in concrete was replaced by another binder, Portland cement, so named because of its supposed resemblance to Portland (a town in England) stone. In modern times, cement-based plasters, known as cement stuccos, have emerged. Cement stuccos consist of a mix of Portland cement, sand, and (frequently) lime, and are often used as finishes on cement or concrete block walls or even wood-framed homes. Although we don't recommend this practice, cement stuccos are sometimes used for finishing natural buildings, including straw bale and adobe homes. Over the years, serious problems have arisen in straw bale and earthen buildings finished with cement stucco.

Cement is often loved by architects and engineers, even some of the world's best: they respect this material and depend on it for the buildings they produce. It was only natural that builders should decide to apply it on earthen and straw bale buildings. However, as a wall finish for natural buildings, cement seems to have exceeded its useful range of application.

In a effort to create a durable, long-lasting finish, manufacturers created synthetic stucco, consisting of latex, sand, and numerous other chemical components, which is used as a surface coating on a variety of buildings or a color coat on cement. Like cement stucco, it has limited value, for reasons that will become clear shortly. Traditional finishes, such as those made of clay-rich soil, lime, and gypsum, are proving to be the superior products for natural buildings.

A Resurgence of Traditional Plasters

Despite the current popularity of cement and synthetic stuccos, earthen plasters continue to be applied throughout many parts of the world. In modern industrial countries such as Canada, Australia, and the United States, there has been a welcome resurgence of earthen plaster in conjunction with natural building, which is proving to be a better option than cement and synthetic stucco in many applications.

The success of these materials in natural building illustrates how ancient techniques can be used today to create environmentally sustainable shelter. Nonetheless, we find that there is a cultural bias toward newer materials. This bias is fueled in part by special interest groups like cement manufacturers who have convinced code officials of the superiority of their products. What better way is there to ensure a good market than to lobby so that your product becomes a required standard?

The resurgence of natural building materials — both plasters and wall materials — and of natural building techniques can't be stopped. A critical number of people have found that, despite cultural bias and building codes that often call for new "state of the art" building materials, natural building materials offer many advantages over modern products. Straw bale and other natural homes are quieter, more energy efficient, and substantially more solid than the best-built stick-frame homes. Natural plasters are not only more pleasant to look at, to touch, and easier to work with, they're better suited to the walls they cover. Bottom line: they feel more comforting and more nurturing.

Today, modern pioneers on the building frontier are finding that they can achieve a balance between our historical roots and current technology. By combining the best of both worlds (for example solar panels for generating electricity and energy-efficient appliances), in homes built of straw bales, rammed earth, adobe, cob, and other traditional building techniques, they're creating shelter that can truly contribute to a more enduring human presence. We now have an opportunity to advance building technologies as well as to promote the cause of sustainability. By using newer technologies and some modern building practices in conjunction with traditional building materials and techniques, we're achieving results that are not just good for the planet, but good for people, too.

2-5: Solar electric panels on this home built from recycled and earthen materials and straw bales demonstrate how old and new techniques and technologies can be used to create more environmentally sustainable houses.

DAN CHIRAS

Why Use an Earthen Plaster?

Some may question your decision to use an earthen plaster in this day of space-age building materials. Why would anyone want to use mud to coat a wall?

Earth Plasters Protect Walls from Wind, Rain, and Fire

Earthen plasters serve a variety of purposes in straw bale and other natural homes. They protect walls from moisture (we'll describe how at the end of this section) and from the erosive forces of wind. They also reduce air infiltration, preventing energy loss and uncomfortable drafts: without a plaster coating, wind would easily penetrate the walls of many structures, especially straw bale homes. Earthen plasters also impede the spread of fire and serve as an effective barrier against pests, among them insects, mice, and other rodents.

Earth Plasters Add Unrivaled Beauty to a Home

Further adding to their value, the beautiful colors and rich, sensuous textures of earthen plasters contribute a unique feel to a home. Because clays occur naturally in many different colors, your choices for room and house color are many and varied. Yellows, reds, tans, browns, and greens are a few of the many colors you'll find in nature. Colored clays and mineral pigments can be used to tint finish coats or to make natural paints (such as alises or clay "paints") that can be applied to an earthen plaster wall (see color gallery). Harvesting local colored clays can be a fun and fulfilling experience. You can also purchase colored clays at local pottery supply stores. The creative options are endless — your imagination is your only limitation.

If worked properly, finish coats take on a soft, inviting look. Some mudders rub a mixture of oil and beeswax on earthen finished interior walls, which produces a lustrous appearance. Mica flakes can be mixed into finish coats to give wall a subtle, but inviting sparkle. Earthen plaster can also be used in conjunction with any form of natural building, including straw bale, adobe, cob, rammed earth, straw-clay, and wattle-and-daub. The interiors of Earthships and other homes built of tires packed with dirt can also be successfully finished with earthen plasters.

Although earthen plasters generally don't adhere very well to large, flat, nonporous surfaces, such as oriented strand board (OSB) or plywood, interior applications may work, provided special steps are taken to create a good adhesive surface (described in Chapter 3). Earthen plaster can be applied successfully to drywall; in Germany, this has become quite the fad! Earthen plaster can even be applied to interior cement stucco walls, transforming the

2-6: Pigmented natural clay finishes applied to these earth plastered walls make this natural earth and straw bale home warm and inviting, a far cry from many modern homes.

CEDAR ROSE GUELBERTH

> ## EARTHEN PLASTER OVER DRYWALL
>
> When applying an earthen plaster to drywall or plasterboard, be sure to apply an adhesion coat consisting of flour paste, sand, and manure first, to provide a roughened surface that earthen plaster will key into. Some folks attach chicken wire or glue burlap to drywall, then apply thin layers of earthen plaster. The results are stunning and will change the feel of a room dramatically.

cold and impersonal cement finish. It provides a warm, soft feel to a space and improves acoustics; the transformation can be remarkable.

Overall, earthen plaster performs extremely well, especially if a house has been designed with care, a subject we discuss in detail in Chapter 3. There are a host of additional factors to justify your selection of a mud plaster.

Mud Plasters are Fun to Work With

Earthen plasters are easy to work with and fun to mix and apply. "Once you've put your hands in that mud mix you don't feel like doing any other type of plaster," write Paul Lacinski and Michel Bergeron in their book, *Serious Straw Bale*. For adults, working with earthen plaster seems like kid's play. For children, it is play! You can't find a more people-friendly material than earthen plaster.

Earthen Plasters are Safe to Work With

The pure joy of earthen plaster ought to be sufficient justification, but there are many other reasons why one would choose this approach over others. Unlike cement-based stuccos and lime-sand plaster, earthen plaster contains no caustic chemicals to eat away at the skin on your hands or to burn your eyes. In fact, after a day of applying earthen plaster, the skin on your hands feels soft and supple.

Earthen Plaster Allows One to Work at a Relaxed Pace

Earthen plasters make for a relaxed work pace, too. You won't find yourself hurrying because the plaster is threatening to set up on you as in a gypsum plaster or cement stucco. In addition, unlike cement stucco and gypsum plaster, you don't have to use up a mix as soon as it is made. In fact, you can cover a mix at the end of the day and use it the next day — or, if the air is cool, within a couple of days. (Some components may go bad if the mix sits around for too long.) If a mix begins to dry out, just add a little water, work it in, then get back to work.

2-7: Earthen plasters are easy to mix and apply and they are lots of fun to work with.

DAN CHIRAS

Earthen Plaster is Easy to Clean Up After

Earthen plasters wash out of clothes and off of skin and out of hair quickly with a little water. They won't ruin tools and equipment as a lime, gypsum, or cement-based stucco will. A little water and a little good old fashioned elbow grease will easily remove plaster from trowels, wheel barrows, and cement mixers — even weeks or years after the plaster has dried.

And it's Recyclable, too

Recyclability and ease of removal are additional benefits of earthen plaster. If you don't like an earthen plaster you've applied, you can scrape it off while its still wet, then try again. If the plaster has dried and hardened, it can be removed with the claw of a hammer, wetted, and reapplied. That's not possible with a cement stucco or lime and gypsum plaster.

Earthen Plaster is Easily Repairable

2-8: Sculpting walls is easy. An earthen plaster is first applied to the wall. A thicker earthen plaster is then used to sculpt forms, as seen in this humorous "baca" by Carol Lee Pelos.

Repair is easier with earthen plasters than with any other form of plaster, especially cement stucco and lime plaster. Cracks or wear in an earthen plaster can be sealed and resurfaced, quickly and effortlessly, hiding all evidence of the damage. Cement stuccos and lime plasters are much more difficult to repair. In cement stuccos, cracks or damaged areas usually have to be chiseled open and feathered so new stucco can be applied as smoothly as possible. Even so, it is hard to create a seamless repair. Because earthen plasters are so easy to work with, they make it easier to remodel or add on to a home.

Earthen Plasters Permit Artistic Expression

For the artists among us, earthen plasters can be sculpted and carved to create intriguing and inviting patterns in wall surfaces. Animals or caricatures can be sculpted on a wall surface, adding an artistic quality or a bit of humor to our homes. When mixed with additional straw and sand, earthen plaster can be used to build out shelves or eyebrows over windows. The cat on the cover of this book is an excellent example.

Earthen Plasters Create a Softer, Quieter Finish

Unlike cement stucco and lime, earthen plasters create a soft, nurturing finish, which is pleasant, even sensuous, to the

JOHNNY WEISS— SOLAR ENERGY INTERNATIONAL

touch. Earthen plasters also create a quieter interior and are acoustically superior to the harder finishes such as cement stucco, creating a more comfortable ambiance.

Earthen Plasters Offer Many Environmental Benefits

Environmentally, earthen plasters are superior to other products. Typically made from locally available materials, they can be harvested with little impact on the environment. You may be able to use the dirt (subsoil) extracted from your construction site — for example, when the foundation or basement is excavated. Alternatively, you may find a seam of clay-rich subsoil nearby. When you've extracted the dirt, you can easily regrade and replant the site or convert the hole to a pond for frogs or a place for deer to enjoy a cool drink.

Locally harvested materials also have a low embodied energy. Embodied energy is the energy that goes into harvesting, manufacturing, and transporting a building material — any building material. Because clay-rich soil can often be harvested onsite with the energy from a couple of granola bars, earthen plasters have an extremely low embodied energy. In fact, earthen plasters made from materials harvested from the building site have the lowest embodied energy of all building materials known to humankind. Clay-rich soil delivered from a nearby source — a neighboring field, for example — has a slightly higher embodied energy than soil harvested on site, but it is still low compared to cement, lime, and gypsum, which have significantly higher embodied energy.

2-9: If it contains enough clay, soil from excavating the foundation can be used to produce earthen plaster for a home.

CEDAR ROSE GUELBERTH

Earthen Plasters are Inexpensive

Another advantage of earthen plasters is that they are cheap. If the subsoil in your area is suitable for making plaster, costs are negligible. The only significant cost is for the labor required to mix and apply the plaster to walls — but you will incur these costs no matter what finish you use. Costs rise, however, if the soil needs to be transported from other sites. In some cases, you may need to purchase powered clay or additives to increase the quality of a mix, which can add to your costs.

Another cost savings comes from the fact that earthen plasters are best applied directly on straw or earthen walls without stucco lath or chicken wire. This reduces labor, resource use, and overall costs.

Earth Plaster: Protecting Natural Buildings from Moisture

Before we move on, we'd like to explore one of the most important benefits of earthen plaster in greater detail — its protective qualities. As we noted above, earthen plasters effectively protect straw bales and other natural materials in the walls of a home from moisture. If a house is designed and built well and you mix and apply plasters correctly, you'll end up with a stable, long-lasting and protective finish.

Protection is provided by four properties of earthen plaster, which result primarily from the presence of clay in a mud plaster. They are water resistance, vapor permeability, rapid drying, and the hydrophilic nature of clay. Let's examine each one individually.

Earthen Plasters Resist Water Penetration

First, earthen plasters are water resistant. That means that water that lands on a wall — for example, in a rain storm — is unable to penetrate the finish. Sure, it will moisten the surface of an earthen plaster, but it generally goes no further. After the storm is over, the moisture evaporates, leaving the wall as good as new.

Please note: We choose our words very carefully in this discussion. When we say that earthen plaster is water resistant, we do not mean it is water-proof like a rubber boot or a plastic rain coat. Water resistant means that it resists water movement through it. The surface layer, when wet, tends to retard any further water penetration. What makes an earthen plaster so water resistant?

The answer lies in the molecular properties of clay. According to soil scientists, clay consists of tiny flat plates. When moisture comes into contact with an earthen plaster, it binds to clay molecules and forms a bridge between them, causing the clay to expand. The binding of water molecules to clay also prevents water from migrating into the deeper layers: the plaster self-seals. This, in turn, prevents water from penetrating any further. Clay's self-sealing properties explain why it is used to line manmade pools, ponds, landfills, and even hazardous waste dumps.

Unlike cement-based stuccos, which wick moisture via capillary action into the interior of a wall, earthen plasters prevent moisture from penetrating into a wall. Thanks to this ability to self seal, clay-based plasters provide a magnificently protective layer over natural materials, preventing damage to the interior of a natural wall system.

2-10: Clay molecules form tiny plates that slide over one another. Water molecules bind to the surface of the clay platelets, causing clay to expand, but also resisting further penetration — thus creating a water-tight barrier.

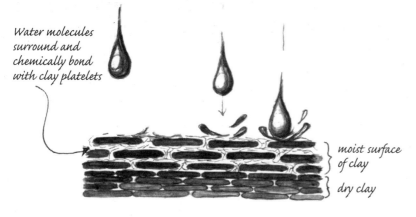

Water molecules surround and chemically bond with clay platelets

} *moist surface of clay*

} *dry clay*

Over a period of time, if unprotected, the finished surface of an earthen plaster may erode as a result of exposure to weather, especially driving rain, wind, and temperature changes. Even so, the plaster in the deeper layers of earthen plaster will continue to protect the wall — in large part, as a result of the clay content. An occasional simple refinishing (resurfacing) with earthen plaster or a clay-rich material called alis will restore the finish, ensuring many additional years of protection. (We discuss earthen plaster maintenance in Chapter 7).

Earthen Plasters are Permeable to Water Vapor

Although it seals itself to liquid water, clay is permeable to moisture vapor — that is, it allows moisture suspended in air to pass through. It's a little like Goretex™ fabric in this regard: Goretex™ stops liquid water from penetrating clothing but permits water vapor to pass out.

In a house in colder climates, water vapor from interior sources (such as showers) tends to flow into straw bale walls primarily through cracks around windows, door frames, or at the junctions of walls and ceilings. In warmer climates, moisture tends to move inward. A small amount, some say about ten percent, flows directly through the wall via diffusion and as a result of pressure gradients.

Water vapor is just moisture suspended in the air inside a house and outside. In most climates, water concentration is higher inside a house than the air outside. Thus, moisture tends to flow (diffuse) out through cracks in walls and, to a lesser extent,

> ### WATERPROOF VS. WATER-RESISTANT: WHAT'S IN A NAME?
>
> When we say that clays in earthen plasters provide water resistance, we mean that they tend to impair water movement into the plaster. They are not, however, waterproof. An earthen plaster will absorb some water and could be washed away by a stream of water. But when wetted, earthen plasters containing clay tend to retard the movement of water past the surface. That is, they resist further penetration of water. That is what we mean when we say they are water resistant, but not waterproof.

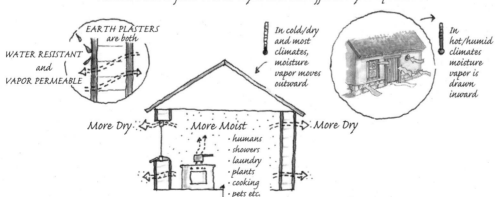

Moisture moves from Wet to Dry and is also affected by temperature

EARTH PLASTERS are both

WATER RESISTANT and VAPOR PERMEABLE

In cold/dry and most climates, moisture vapor moves outward

In hot/humid climates moisture vapor is drawn inward

More Dry ⇦ More Moist ⇨ More Dry

· humans
· showers
· laundry
· plants
· cooking
· pets etc.

Sources of Indoor moisture

2-11: Air containing moisture moves into and out of the walls of a house primarily through cracks around windows and doors and between the foundation and walls.

VAPOR PERMEABILITY

Both earth and lime plasters bond well to earthen walls, and both are permeable to water vapor. This is important because it allows any moisture absorbed by the walls to escape harmlessly through evaporation. Lime and earth plasters are said to "breathe."

— Michael G. Smith, "Earth Buildings and Moisture," *The Last Straw*

---

through the wall itself. (Diffusion is the movement of water vapor from an area of high concentration to low concentration.) Just the opposite occurs in hot, humid climates.

Water vapor moves in response to pressure gradients. In most instances, air pressure inside a home is higher than outside. This tends to cause air to move outward. If the air contains suspended water molecules, they travel outward as well.

In a straw bale home finished with cement-based stuccos, moisture enters walls through cracks and also moves by capillary action through the cement stucco finish into the interior of a wall. However, problems soon arise, for water vapor tends to get trapped inside the walls. It often condenses on cooler surfaces inside walls, too. Over time, water can saturate the straw bales in a home with cement stucco walls. Mold and mildew will very likely form and the straw will begin to rot. Wood in walls — for example, around windows or in posts and beams — can mold and rot as well, which may lead to serious health problems as well as structural problems. Earthen plasters, on the other hand, allow water that enters a wall to diffuse out, protecting straw, wood, and earthen materials from deteriorating.

Clay plasters protect walls in other ways, too. For example, earthen plasters on interior walls tend to absorb moisture in the indoor air. If moisture levels inside a house fall, as they will on a warm sunny day or when a house is unoccupied, the moisture in the plaster will diffuse back into the room air. The plaster, in other words, dries out before the moisture can make its way to the interior of the wall.

2-12: (a) Water penetrating a cement stucco is often trapped inside the wall, causing damage. (b) In a straw bale wall finished with an earthen plaster, moisture that enters the wall as water vapor can exit through the vapor permeable plaster.

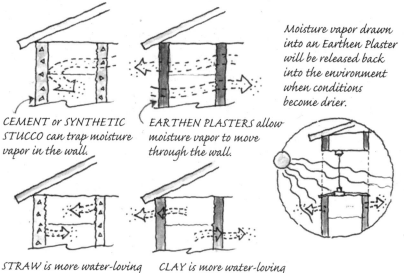

Moisture vapor drawn into an Earthen Plaster will be released back into the environment when conditions become drier.

CEMENT or SYNTHETIC STUCCO can trap moisture vapor in the wall.

EARTHEN PLASTERS allow moisture vapor to move through the wall.

STRAW is more water-loving than CEMENT – thus straw draws moisture vapor from the cement.

CLAY is more water-loving than STRAW – thus drawing moisture vapor from the straw!

OF WATER AND WALLS:
UNDERSTANDING THE MOVEMENT OF WATER THROUGH WALLS

Water is one of the most detrimental environmental forces acting on a house — any house! — and can cause serious damage, even collapse of the structure. Understanding water movement through foundations and walls helps us design better homes and it helps us select proper wall finishes and building strategies.

As a rule, water vapor flows from warm to cool areas. Because warmer air holds more moisture than cooler air, the warm-to-cool flow of water vapor really is a flow from high water vapor concentration to lower water vapor concentration.

In most climates, water vapor tends to move from the inside of a house (warmer, moister environment) to the outside (cooler, drier). As noted in the text, water also tends to move down pressure gradients. In many cases, air inside a house has slightly higher pressure than the air outside. This too causes water vapor to move from the moist interior to the drier exterior. But where does the water inside a home come from?

Inside a home there are many sources of moisture. Humans and pets produce gallons of moisture from perspiration and exhalation. Stoves, showers, plants, washing machines, and dishwashers generate moisture. Water vapor penetrates walls through inevitable cracks and openings in the building envelope — for example, around windows or light switches. It also penetrates directly through wall surfaces, although to a much lesser degree.

In conventional homes, water moving from a warm to a cooler area tends to condense when the temperature inside the wall reaches the dew point — the temperature at which water vapor condenses. In New England, for instance, water vapor escaping from the interior of a home condenses on the inside surface of the outside wall, where it collects, causing damage. Vapor barriers in conventional homes prevent this from happening.

In natural homes, vapor barriers are not required — in fact, they can cause extensive damage to a wall made of natural materials. "Breathable" plaster is required to prevent moisture from condensing inside walls where it could promote the growth of mold and mildew, leading to decay of straw and wood components of a wall system, or deterioration of earthen materials in adobe or cob walls. That is, you need a plaster that permits water vapor to pass through.

We recommend a vapor permeable plaster on both sides of a wall. For exterior walls, earthen and lime plasters are suitable. For interior plasters, add gypsum to the list. Any combination of them will work well.

Earthen Plasters Dry Rapidly, Which Helps Protect Walls from Moisture

Formal and informal studies show that clay tends to dry out rather rapidly. According to studies in Germany, clay releases water vapor more quickly than cement and other building materials. As a result, water splashed onto a clay plaster wall will evaporate more quickly than water splashed on a cement stucco wall. Consequently, there's less chance of water penetrating the interior of the wall.

HYDROPHILIC: WATER-LOVING

If you have studied chemistry, you have undoubtedly run across the term hydrophilic which is used to describe substances that attract water. Clay is one of those substances: it has a natural affinity for moisture. In an earthen plaster, clay draws water away from bales, thus protecting them.

Clay's Water-Loving Nature Helps Protect Walls

Clay tends to draw moisture away from straw bales — that is, it draws moisture away from straw and other materials in a wall, such as wood. Dry clay, in fact, is a natural preservative in archeological sites.

Because water is required for the decay of organic materials and because clay is more hydrophilic (water-loving) than other building materials like straw, clay in plaster acts as a preservative by sucking moisture away from straw walls (see sidebar). Cement, on the other hand, is much less hydrophilic than clay. In a straw bale wall, in fact, straw is more hydrophilic than cement and therefore it tends to draw moisture from cement stucco. Moisture inside bales can promote mold and mildew and rotting, causing an earthen wall system to deteriorate.

Some of the Downsides of Earthen Plasters

We're strong advocates of earthen plasters. As the previous list of benefits illustrates, earthen plasters are ideally suited for natural homes, as well as the people who build them and the folks that live in them. As with anything, there are disadvantages to earthen plaster, and we offer our perspective on these, too. We believe that you should go into this venture aware of any potential problems — and ways to avoid them.

Earth plastering requires a lot of work, but so does any plastering job

Some people complain that earthen plasters are labor intensive. We agree, but add that so are all forms of plaster or stucco. "Whether you apply (a plaster) by hand or by machine, the sheer physical effort required to plaster a building surprises many people," note Chris Magwood and Peter Mack, authors of *Straw Bale Building*. "Be prepared to put in a lot of effort yourself or to pay well for the people you hire to make the effort for you," they add. But remember, too, if you organize your work site and find ways to mix batches quickly and efficiently (a topic we'll cover in Chapters 5 and 6), earth plastering can be efficient, fast, and fun. Earth plaster goes on quickly and easily. It's a lot less work than cement stucco or lime plaster and a heck of a lot more pleasant to work with.

Earth Plaster Requires Some Maintenance, But Much Less Than Other Forms of Plaster

Earthen plaster may require some maintenance. As a rule, however, if your plaster is mixed and applied well, and the building is designed correctly, earthen plasters will require less maintenance than a cement stucco. This is especially true in climates subject to freezing temperatures, where repeated freezing and thawing of water that has wicked into cement stuccoed walls tends to cause cracking.

Buildings also move — that is, shift a little — over time, which also causes cracking. Cement stucco isn't very flexible: earthen plasters have more give. Remember, too, that repairs in earthen plasters are a heck of a lot easier than in lime-sand plaster and cement stucco. It is easy to wet down an area, then replaster it. And the results are much better: you'll hardly notice the repair work, if at all. You can't say that about many other forms of plaster! Maintenance is a breeze with earthen plasters. Occasional touch-ups to worn areas and refinishing (using plaster and alis) are all that are needed.

Earthen Plasters Can Dust — If They're Not Prepared Correctly

Another potential downside is that earthen plaster finish coats may "dust," especially when a person or a pet brushes against it. To this we note that dusting is generally a problem only when the finish coat was not properly made. As we'll explain in Chapter 6, be sure to run test patches to see if your mix will dust. If the test patch creates dust, some natural additives usually remedy the problem.

If a finish coat that you've applied is dusting, there are ways to fix it. You can, for instance, apply an alis (clay paint) to seal the surface of wall, eliminating the problem. You can also apply a clear casein paint or solution of potassium silicate. We'll describe how to make a dust-free finish coat in Chapter 6, clay and silica paints in Chapter 7.

Earthen Plasters May Not Be Approved by Local Building Codes

Another concern is that earthen plasters may not be approved by local building codes. Moreover, much to our chagrin, some building codes actually require the use of cement stabilizers or asphalt emulsion in earthen plasters, requirements we think are unnecessary and potentially detrimental to human health and the structural integrity and performance of the plaster. Asphalt emulsion is a toxic substance that shouldn't be in any wall. Other codes may require that you apply earthen plasters over cement stucco. Earthen plaster over cement stucco will not function well. Cement stucco significantly reduces its breathability and will not hold up in exterior applications. For interior walls, building codes may require earthen plaster to be applied over a gypsum base coat.

These potentially problematic requirements result from the fact that, for many building departments, officials rely on rules established for standard building systems. In an effort to provide regulations to protect the health and safety of homeowners in their jurisdictions, building department officials tend to apply what they know to be true for standard building systems. Their lack of experience with natural building materials can lead to some detrimental requirements for natural homes. These can lead to serious problems in natural buildings. We urge you to work with local officials to promote a better understanding of earthen plasters. We will address building codes later in this chapter.

You Need to Know What You're Doing

As with any construction technique, you need to know what you're doing. Educating yourself on earthen plasters and alternative construction techniques is extremely important. Reading this book will help; however, we recommend hands-on training, too. Although there aren't many publications on earthen plaster, the Resource Guide at the end of the book lists what is available as well as workshops where you can hone your skills.

Lack of Local Experts

Another potential problem is the lack of local skilled earthen plasterers. Although there may be many people skilled in cement and synthetic stucco in your area, good earth plasterers are few and far between. You may need to travel out of your area for workshops and training. When building a house, you may need to hire someone to consult for a few days, which means obtaining advice over the phone or hiring someone to travel to your work site. If you want to hire professional plasterers to do the job, you may have to bring a crew in from another city, state, or province.

There's No Precise Formula for Earth Plasters

Because subsoils used to make plaster vary dramatically from one location to another — and may even vary on the same site — we can't give you a precise fail-proof formula or recipe for successful earthen plasters. Unlike cement stucco, which is a fairly standardized material, the proper mix for an earthen plaster is unique to each site. When making earthen plaster you will need to rely on your knowledge, intuition, experience, and experimentation.

Some subsoils may be perfect as they are: all you'll need to do is to add straw. More likely, however, you'll need to doctor your mud a bit, adding not only straw but sand or even clay or flour paste to increase its adhesive properties. But that's what's fun about earthen plaster: it allows us to experiment. If you understand the function of the various components of an earthen plaster and are open to a little experimentation, you can perform some simple tests to figure out the best mix. We'll discuss this topic in more detail in Chapters 4 and 6.

Mixing gypsum and lime plasters and cement stucco are more straightforward than an earthen plaster, because the products generally come out of a bag. Even so, local plasterers may use slightly different mixes to accommodate regional climatic peculiarities. If you don't know what they are, instructions in a book or on a bag could lead to less than optimal results.

Earthen plaster does have some disadvantages, but we think they are few and largely preventable. Moreover, these minor shortcomings are clearly outweighed by the many advantages of this marvelous material. If done right, an earthen-plaster is safe, nurturing, functional, and long-lasting. "If done right" means designing a house properly and building walls correctly, topics we will discuss in Chapter 3. In addition, it means achieving the right ratios of sand, clay, and straw — mixed correctly. And, last but not least, they need to be applied properly. Chapters 5 and 6 will cover these topics.

Are Earthen Plasters Applicable Only in Dry Climates?

Our experience has shown that earthen plasters are suitable in all climates, even in areas with driving rains, if a home is designed correctly and built well. That's true of any plaster — cement, lime or earthen. For example, a woman in Australia built a two-story straw bale home and finished it with cement stucco. One wall was in the direct path of incoming rain and was continually wet, a result of an oversight in the initial planning and design. The owner discovered that rainwater seeped into the interior of her walls, and in short order her straw bale walls began to rot. Had she sealed the wall, say painted it with a latex paint, to prevent water from entering the straw bale interior, she would have had the same problem — rotting bales — this time due to moisture trapped inside the walls, as explained earlier in the chapter.

To ensure the success of a plaster, any kind of plaster, a home needs to be designed to be protected from interior and exterior

> **EARTHEN PLASTER: A FINISH FOR ALL SEASONS AND ALL CLIMATES**
>
> Earthen plasters work well in many climates — if the house is designed and constructed properly, they can be used alone or in conjunction with other types of natural plaster, such as lime. Earthen plasters are not just for hot, dry climates.

moisture, rain, melting snow, and wind. Remember: no matter what finish you're going to use, you'll have trouble with driving rains or snow melt if your home was not designed correctly. We suggest you study carefully the design guidelines we present in the next chapter.

Earthen plaster can be made more weather resistant by adding natural stabilizers, which increase the durability and water resistance of the surface. Your choices of additives vary, depending on your climate and the availability of materials. A few examples of natural stabilizers are cooked flour paste, prickly pear cactus juice, and lime. We'll talk more about these and other additives in Chapters 5 and 6. But remember, no matter how great your mix is, if your building is not designed well, you'll have problems.

Some straw bale builders believe that straw can be added in greater quantity to the finish coat of earthen plaster to increase weather resistance. Experience has shown that a high proportion of straw in the finish coat can be detrimental. Straw tends to provide an avenue for moisture to move into an earthen plaster, for example, when rain moistens a wall. Moisture drawn into walls can do considerable damage. Straw in the finish coat may cause moisture to run down walls, eroding the clay in the finish plaster. A straw-free or minimal-straw finish coat creates a sealed, bonded finish that resists erosion. Therefore, while high straw content in a finish coat can work in some climates, especially dry ones, in most climates it can be problematic. Bottom line: we recommend against the use of straw in exterior finish coats.

In harsh climates, take care to design your home to protect the plasters. You may also want to consider applying earthen plaster as a base coat for a lime plaster finish coat. Lime plaster is an excellent finish coat that provides great protection against the elements and is extremely compatible with earthen plasters.

Why Not Use a Cement Stucco?

Cement stucco is widely available and commonly used in straw bale homes throughout the world. It has also been used on cob structures as well as adobe homes in the United States and other countries. Cement stuccos can be applied by hand or by stucco gun, which greatly accelerates the process. Building departments may favor or require cement stucco, not just on straw bale homes, but earthen structures such as adobe and cob.

Cement stucco is considered the best "plaster" option in cold, wet climates, by many straw balers — an assertion to which we respectfully take exception. In fact, we think that the use of cement stucco on straw bale or on virtually any other natural building material represents a colossal mistake that will come back to haunt people in the years to come. As we pointed out earlier, and will do again and again, cement stucco wicks moisture into the interior of walls. It cracks and lets moisture in as well. Cement stucco traps moisture. Moisture in straw bale and earthen walls causes them to deteriorate.

Cement stucco also prevents moisture trapped inside a wall from escaping, which could cause considerable damage to straw bales or other earthen materials. Writing in *The Last Straw*, Michael G. Smith summed up the situation in an earthen wall: "Unable to evaporate through the cement stucco, this moisture accumulates over time, saturating and weakening the wall, especially at the point where earth meets cement. The wet earth turns into mud and flows, leaving invisible cavities behind the plaster." Smith goes on to tell about many old adobe buildings in New Mexico that have been plastered with cement stucco with disastrous results. "The St. Francis Church in Rancho de Taos, New Mexico, built in 1815 with massive adobe walls and thick buttresses, was plastered with

2-13: The combination of poor design details and cement stucco leave this straw bale structure susceptible to moisture wicking up the wall and into the straw bales, which can cause severe damage to the underlying straw bale wall system.

cement stucco in 1967," he notes. "In 1978, water trapped by the cement was found to have eroded the wall to a depth of up to two feet." As a result, the stucco had to be removed and much of the church had to be rebuilt. Traditional earth plastering has now resumed, replacing this well-intentioned but misguided effort. Smith goes on to say, "Similar stories may be told about centuries-old cob buildings in the British Isles, which have suffered severe water damage after being stuccoed with cement or having nonbreathable paints or wall paper applied to interior surfaces." Despite these and similar experiences, New Mexico's building code required new adobe homes to be finished with cement stucco until 1998, when the requirement was quietly dropped.

Cement stucco can be equally damaging to straw bale buildings. The story we told earlier about the disaster in an Australian straw bale home is a good example.

Another reason why we don't recommend cement stucco is that cement is a high-embodied-energy material, meaning that it takes a great deal of energy to mine, manufacture, and ship. It is also a pain in the neck to work with — rough on hands, causing skin to crack, and hard to clean up after use. Cement-based stucco may even eat away at straw and natural materials. If it sets up in a wheelbarrow or on tools, forget it. Dried cement stucco is very difficult to remove!

Perhaps even more important, cement stucco doesn't adhere well to straw or earthen materials. In both instances, it is applied on a chicken-wire mesh that's tightly

CEMENT STUCCO VS. EARTHEN PLASTER

The importance of earthen plaster's permeability to water vapor became evident in the adobe St. Francis de Assisi Church in Ranchos de Taos, New Mexico, which was built in the early 1800s. Originally earth plastered, the church was coated externally with cement stucco in 1967 in an attempt to reduce annual maintenance — that is, periodic resurfacing of the church with a protective alis (a clay-based finish coat). Unfortunately, as locals soon discovered, the cement stucco wicked moisture into the interior of the walls. Cracks in the stucco also permitted moisture to enter, where much of it was trapped, softening the adobe and causing considerable deterioration. The cement stucco has since been removed and the old practice of surfacing the walls with clay has been resumed. (See color gallery).

2-14: Chicken wire, mesh, or metal lath over straw bales often creates air pockets (as shown here) that can weaken the plaster.

CEDAR ROSE GUELBERTH

2-15: Cement stucco may crack or flake off, permitting water to penetrate the walls. This home was completed only a couple of weeks before photo was taken.

DAN CHIRAS

attached to the straw, which takes more time and costs more money. Some builders are now using larger 14-gauge square welded mesh (2" by 2" mesh) instead of chicken wire. Either way, adding mesh increases the cost of plastering by adding materials and labor. Wire mesh prevents stuccos from locking or keying into straw bales. It also creates air pockets that weakens the plaster.

Cement-based stuccos also form nasty cold joints between one day's work and the next. Special care has to be taken to blend a previous day's work into the present-day's work. Earthen plasters, on the other hand, repairs easily with little sign of your handiwork.

Cement stucco also cracks, as noted earlier. Unsightly cracks often occur around windows and doors, and may also appear along walls where settling occurs or where there is exposure to hot sun and cold temperatures. While careful building practices can reduce cracking, it is still pretty hard to avoid. Besides being unsightly, cracks need to be repaired quickly to prevent moisture from entering and further widening or lengthening of the crack.

Cement-based stuccos also wick moisture, which is then subject to the freeze-thaw cycle in cold climates. Successive freezing and thawing of moisture in cement stucco causes it to crack more, which in turn lets in more moisture. Entire sections may flake off, a process referred to as spalling.

Finally, cement-based stuccos release chemicals that are often added to the material during the manufacturing process in order to retard drying or improve the stucco's workability. These may have adverse health effects.

Codes and Earthen Plaster

Building codes for straw bale construction are few and far between. Those that do exist generally permit the use of earthen plaster for interior and exterior walls. It can be applied directly to bales without reinforcement, but codes relating to weather-exposed exterior walls almost universally stipulate that earthen-plastered walls need "to be stabilized using a method approved by the building official." They may require lime, cement, or asphalt emulsion additives to stabilize — increase the weather resistance of — your plaster.

According to David Eisenberg, director of the Development Center for Appropriate Technology, "It is possible to use earthen plasters, both stabilized and unstabilized in some jurisdictions, but it is often dependent on getting special approval from the building official. Generally, the more weather-protected the wall, the more likely that you will be allowed to use unstabilized earthen plasters. That is, the more your design protects the walls from exposure to the elements, especially rain, the greater your chances of getting approval for an earthen plaster. Where there is direct weather exposure, many jurisdictions or building officials will require some sort of stabilization of earthen plasters."

According to Eisenberg, the codes for adobe and rammed earth are a bit different. "In most cases," he notes, "codes state that stabilized materials do not need to be finished with exterior plaster, but unstablized earthen walls must be protected by cement stucco — even though there is ample evidence of the problems of applying cement stucco over unstabilized earthen walls." So, if you've added cement or some other stabilizer to your adobe or rammed earth wall, you're not required to apply a coat of cement stucco. If you haven't, you need to apply a cement stucco — which is a dangerous and foolhardy proposition for the reasons noted above.

What if the building code requires a cement stucco on straw bale or other natural homes?

Despite their potentially devastating weaknesses, cement stucco finishes continue to be used, even recommended for certain climates. Our advice on the matter: don't do it!

Build a respectful and cooperative relationship with your building department. You will be in a better position to explain why an earthen plaster should work on your home. To aid in your effort, you may want to provide building department officials with educational materials about the durability of earthen plasters and the downsides of cement stuccos. A copy of this book might help. Contact the folks at *The Last Straw* for documentation on troubles caused by applying cement stuccos over straw and earthen building materials such as adobe. There are a lot of good examples.

Another approach is not to make a big deal about the plaster you're using. You can spec your finish as a "mineralized fiber stucco." We've found that building department inspectors scrutinize many aspects of a project during onsite inspections but often do not pay any attention to the type of plaster you're using. If you build a good relationship with inspectors and building code officials and are building a solid structure and have taken into consideration moisture issues, they may trust your choices. Or you can try to avoid attention, although we can't officially recommend that you subvert the law.

Another option is to specify a lime-sand plaster finish coat over earthen plaster. If all else fails, go with a cement stucco, but apply the elements of design and construction we describe in Chapter 3.

Can I Use a Synthetic Stucco?

Synthetic stucco is a popular plaster on modern buildings. It is typically applied over cement stucco or on foam applied to exterior sheathing (OSB) of stick-frame houses. It is the product you often see in the fake adobe homes that are popping up throughout the southwestern United States. Synthetic stucco is also occasionally used on straw bale structures. This is not a good option for a plaster on a straw bale or earthen wall. Why?

Synthetic stucco is made from pigment, latex, sand, and other chemicals. When it dries, it forms a relatively impermeable layer that prevents moisture from passing through a wall. It is, in other words, a non-breathable wall finish. If moisture enters walls through cracks, its outward movement, already retarded by cement stucco, will be further impaired by synthetic stucco.

Synthetic stucco is also fairly expensive. Moreover, it releases potentially harmful chemicals that could cause health problems. It has an odd, synthetic look, too. Needless to say, we don't recommend it for use on straw bale homes or earthen homes.

Quick Recap

Plasters have been used for centuries to protect walls of homes and other buildings. The earliest natural plasters were made from mud and are known as mud or earthen plasters.

Over time, builders began to discover other suitable wall finishes. Locally available gypsum, for instance, could be transformed into a plaster for interior uses. Lime plasters could be made from processed limestone. In more modern times, cement and synthetic wall finishes, known as stuccos, emerged.

Today, many builders are turning to older materials, especially earthen plasters for natural homes. Earthen plasters are unequivocally good for the environment. The three primary components — earth, sand, and fiber — are usually locally available or available nearby. Their harvest results in little impact on the environment and damage can be easily restored. Earthen plasters also have low embodied energy. The energy from a few granola bars is all that's needed to extract the stuff. Very little, if any, fossil fuel needs to be consumed, unless you use a backhoe or transport materials in from other sites.

Cost is another strong consideration. By and large, the components of an earthen plaster cost very little — or nothing at all. If you're able to extract the clay-rich soil from your building site and it contains the right amount of sand, your raw materials are free of charge. If you need to extract dirt from a nearby site, you may have to pay to have them transported to your site.

Earthen plasters do not require the application of chicken wire over the straw bales. They actually perform best without chicken wire or metal lath. Earthen plasters key straight into the bales, eliminating the potential for air pockets. The result is a very solid finish.

If appropriately mixed and applied, earthen plasters are extremely durable. In addition, they are easy to repair and maintain.

For natural buildings, one of the most important advantages of earthen plasters is that they are breathable. That is to say, they permit moisture (water vapor) to pass through. At the same time, they are remarkably resistant to water that lands on walls — for example, during a rainstorm.

Earthen plasters are easy to mix and apply. They're lots of fun to work with, too. When it comes down to it, earthen plasters feel good to work with. They can be very forgiving as well. And, of course, they add an unparalleled beauty to a home while connecting us with our roots, the soil from which all life springs. Earthen plasters not only perform well and are user-friendly, they are relatively easy to master.

Earthen plasters don't rely on chemical additives, new technology, or new techniques for application, although a wheelbarrow and a cement mixer will come in handy. Like other natural building materials, they have remained unchanged for centuries. Clay-earth, sand, fiber, and water remain the primary components.

Despite the diversity of approaches, all earth plastering relies on the same components and concepts. Adaptations in plaster mixes and application may improve the results. Therefore, in this book, what we hope to convey is an understanding of the properties of the various components of earthen plaster and the functions of each. We hope that this information will help you determine what mixes will work in your area on your project. We will offer guidance on plaster recipes, but by and large you're going to have come up with your own recipes. The feel of an earthen plaster and the way it goes on a wall will be more important than any recipes we could give you. In many ways, our challenge is like that of a cooking teacher: to become a good cook you need to understand food texture, flavors, what happens when ingredients are combined, and various techniques to convert the raw materials that go into a dish into a delectable meal. Plaster is a craft and an art.

CHAPTER 3

Designing and Building for Earthen Plasters

MATTS MYHRMAN, AUTHOR OF *Build It With Bales*, a popular how-to book on straw bale construction that has guided many people through this exciting and challenging process, confided that he has attended "too many wall raisings" at which he asked the homeowner, "How do you plan on finishing this?" after the wall was complete. The response all too often is, "I don't know, maybe I'll try an earthen plaster."

The decision to apply an earthen plaster is a good one, yet in some instances a home's design may make earthen plaster a foolhardy endeavor! Our advice on the subject is this: decide early on — and then design your building for the finish you select, or you'll be destined for problems down the road. We can't say it strongly enough: don't relegate plaster to an afterthought. The key to the success of a plaster, any plaster, is to design a home for it. To understand why, we examine each step in the construction of a home to see how it affects plasters and finishes. With this information, we can design each step in the construction of a natural home to ensure the success of plasters and finishes.

> Earthen plastering actually begins long before a house is built, at the design stage, or at least it should. Design for earthen plasters from the beginning, before you break ground or stick a window in a wall!
>
> — Cedar Rose Guelberth

Designing Homes for Earthen Plasters (or any finishes for that matter!)

For an earthen plaster to be successful on a straw bale or a natural home, careful planning is required. Map out the details well in advance to eliminate design and construction mistakes. As you will soon see, moisture is a key issue in your considerations. In straw bale and natural construction, as in all other types of building, moisture can greatly influence the success of a building, adversely affecting a plaster and its ability to protect a wall.

When designing a house, you will need to look at a number of factors. First is climate and ecology. Next is siting and landscaping. Then comes internal moisture levels. With this information in mind, you can design the foundation, walls, and roof.

Climate and Ecology

When first considering earthen plasters in a design it is important to become familiar with the climate of the area for which you are designing. Year-round weather information, including moisture levels, temperature and humidity levels, onsite weather patterns, and wind patterns are all aspects you'll need to study. How rainy is it? Does the rain fall in a constant drizzle or does it come in driving torrents that will slice into the earthen plaster walls of your home, washing it away in tiny rivulets? What direction do storms come from? Do they come from different directions at different times of the year?

> One of the most common problems with straw bale and earthen homes is that they are designed for Arizona when they are going to be built in Colorado or Louisiana.
>
> — Matts Myhrman

Snow is less of a concern than rain, but don't discount its importance. Before designing a home, you need to know a few facts about snow in your area. For example, how heavily does it snow? Does snow melt quickly, or will it accumulate and build up against the walls of your home? Will wind cause it to drift and accumulate alongside the house? Are there any locations on your site that are more prone to drifting than others? The list of questions goes on.

You will also want to consider more specific weather exposure — how a home will be exposed to and affected by weather at different times of the year. Weather exposure is influenced by various geographic features, such as nearby mountain ranges, hills, and so on. Understanding the way the weather acts upon a home will provide important design information, not only to enhance the effectiveness of plasters but to increase the longevity and energy efficiency of a home. Analyzing the impact of weather on

3-1: When designing a home, you'll need to consider local ecology and year-round weather information, as explained in the text.

a home will help you determine which walls need more protection and what landscaping strategies will help keep moisture away from the building. This information will also assist in the design of drainage systems, roof systems, foundations, and other elements.

Dealing with moisture is a key issue in building natural homes and creating durable plasters, as it is in any kind of construction. Moisture on the site (in the ground, for instance) can easily wick up into foundations or creep up plaster into walls, which then leads to rapid deterioration of walls. It can also enter through other routes, affecting both plasters and the underlying walls. (See Figure 2.13.)

In summary, the interaction of climate with your home design and construction plays an important part in the longevity of a plaster and the overall success of a home. The more you know about this interaction, the better job you can do in the siting, design, and construction of a home. As you identify climate conditions and collect year-round weather information, be sure also to identify the local ecology of the area you are building in, evaluating the plants and trees and the wildlife in the area.

Proper Siting and Landscaping

When evaluating land for a building site, remember that it is paramount to prevent water from reaching the area around the building. This is the first line of defense against moisture; landscaping to shed moisture away from the foundation is the second line of defense.

Proper siting for sustainable living entails many considerations, including access to the sun for passive solar, access to natural building materials, and protection from the elements to increase energy efficiency. Proper siting also requires consideration of weather patterns — and how weather will act upon a house. Be sure to assess prevailing winds, snow build up, solar exposure, and other factors that might affect the house.

When considering where to place a home on a piece of land, study the topography of the site to determine where water will flow in case of rain or after snow begins to melt. You'll also want to consider water flow patterns around a prospective home and how water can be shed away from it. Avoid building in sites where water naturally accumulates, or in the direct path of natural flows. If you can't avoid such placement, you'll need to take steps to divert water away from the house.

Be aware of potential flooding from nearby surface waters, rivers, streams, springs, and pooling or water accumulation. Choose a site away from rivers or streams that could overflow — they can all flood under the right conditions. Especially important in this regard is the need to build out of flood plains — low-lying areas along streams and rivers that periodically flood. Rising waters around a home can damage the foundation and the walls of a home.

In arid climates, watch out for arroyos. These dry washes are very deceptive. Sites that haven't seen a drop of water in a decade can fill with water as a result of an afternoon thunder shower sometimes many miles from a home. A flash flood carrying water, dirt, rocks, and debris can not only saturate a foundation and walls with water, it can also demolish a home, sweeping it off its foundation without a moment's notice.

SITING
CONSIDERATIONS

Climate
Weather patterns
Topography
Moisture sources
Ecology

CEDAR ROSE GUELBERTH

3-2: Shaping the landscape to form these diversion shelves upslope of the building helps to move water away from the house, protecting the foundation, walls, and plaster.

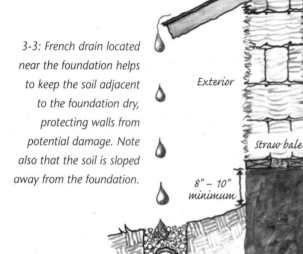

3-3: French drain located near the foundation helps to keep the soil adjacent to the foundation dry, protecting walls from potential damage. Note also that the soil is sloped away from the foundation.

Exterior

Interior

Straw bale

Finished Floor

8" – 10" minimum

Foundation

FRENCH DRAIN slants to daylight or dry well

Landscape and grade a site so that moisture from rains and snowmelt moves away from a house. If you plan to build on a hill or slope, be sure to divert water away from the uphill side of the home, starting considerably above the structure.

Be certain to shed moisture away from the downhill section of the home as well. During construction, be sure you don't create a low spot that attracts moisture buildup on the lower side of your building.

In wetter climates and in certain soil types, water tends to pool up in low-lying areas around homes. It may then seep into the foundation, causing damage to the foundation, natural walls, and earthen plaster. To prevent this problem, be sure to fill such areas, seal the foundation well, and create a good drainage system.

No matter what your site is like, slope the final grade away from your foundation, and provide a French drain on the outside perimeter of your building to carry water away.

Whether on a slope or in the flat lands, you'll have to move a little dirt around to ensure proper drainage. It generally doesn't take much work and the investment in time and money is well worth it, especially considering the potential problems.

Foundation Design

Common sense construction tells us that keeping water away from the foundation of a home — any home — is vital. Water can wick into a foundation and may also wick up into the walls. In a straw bale home, this moisture can cause the straw to mold and rot, resulting in costly damage. Water wicking into an earthen home — a cob or adobe home — can cause the wall to soften and give way. Water wicking up through the foundation may also damage earthen plaster, causing it to deteriorate. In addition, water collecting around the foundation can freeze in cold climates. Freezing can lead to expansion of the soil, foundation heaving, and cracking in the foundation, walls, and finishes.

Remember: siting is the first line of defense against water; shedding water away from a building is the second line of defense. Next comes actual design features of the house itself. What features provides this next line of defense?

A Good Pair of Boots

Ask architects or builders how to protect a house from moisture and they'll tell you, "A house needs a good pair of boots and a good hat to stay dry." We'll concentrate on the good pair of boots — that is, a good foundation — first.

Although you can protect your house from flooding by building on higher ground away from streams and rivers, and from moisture problems by diverting surface water away from the building, or by building on sloped property and grading the site to shed water, some water will enter the soils around your home. To keep this water from wicking into the foundation, you must design a foundation that prevents moisture in the ground from reaching straw bale or earthen walls and their protective coating of earthen plaster.

The foundation of a home not only protects the structure from moisture, it protects the walls from settling and shifting, and it anchors the structure to the ground, preventing it from being blown away or damaged by winds and other natural forces such as earthquakes. All too often "economical" foundations are built by people trying to reduce work and save some money. Jeff Oldham, manager of the design and consulting group at Real Goods, once remarked, "Architects tend to overbuild foundations. Owner-builders tend to underbuild them to save money and time."

Don't skimp on your foundation. For all the heart and soul that you pour into this building, not to mention money, you'll be sorry in the long run if you underbuild the foundation. Although a poorly built foundation might work for a while, over time it may begin to deteriorate, resulting in structural damage ranging from mild to severe. What use is a well-built home if the foundation it rests on is ill-conceived and poorly constructed?

Not much.

How do you design a foundation to protect a home from moisture? No matter which of the many different kinds of foundations you select, it should raise earthen or straw bale walls off the ground at least 8 to 12 inches (20 – 30 cm) at the minimum — more in heavily wooded environments, very wet or severe climates, or on the side of the home from which snow drifts or rain storms come. That way, if water does pool up outside a home, it won't seep into the walls or damage the plaster.

Remember: the height of a foundation above final grade depends on climate and site conditions. All foundations should be well drained to remove any water that reaches

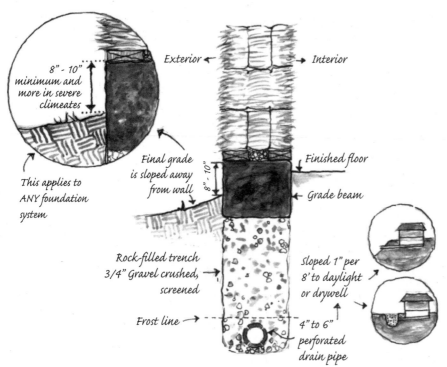

8" - 10" minimum and more in severe climeates

This applies to ANY foundation system

Exterior ← → Interior

Final grade is sloped away from wall

8" - 10"

Finished floor

Grade beam

Rock-filled trench 3/4" Gravel crushed, screened

Sloped 1" per 8' to daylight or drywell

Frost line →

4" to 6" perforated drain pipe

3-4: Rubble trench foundation excavated below frost line is filled with crushed rock. A four- to six-inch perforated drain pipe at the bottom of the trench slopes to daylight or a dry well, removing moisture from the foundation. Grade beam rests on top of the rubble trench and top of grade beam needs to be above final grade, which is sloped away from the house.

them. With this overview in mind, let's take a look at several common foundations — and ways to design them to remain as dry as possible.

RUBBLE TRENCH FOUNDATIONS. Of all of the options, this simple, inexpensive foundation is one of the best from the standpoint of water protection. A rubble trench foundation consists of a trench dug below the frost line, the bottom of which is slightly sloped to ensure adequate water flow to "daylight" or to a sump. As illustrated, a perforated 5" pipe (commonly referred to as a "French drain") is laid in the bottom of the slightly sloped trench (which drains to daylight). The trench is then filled with compacted 3/4-inch screened, crushed rock or pumice, covering the pipe to ground level or grade. A "grade beam" is then formed on top. Grade beams can be constructed out of several types of building materials. The most commonly used are concrete, concrete block, stone, and Earthbags.

To prevent moisture from entering and wicking up into the straw bales or earthen walls, be sure to seal the grade beams and foundation walls with a nontoxic waterproofing material (discussed shortly) on all sides. In addition, exterior finished grade should backfill a few inches up the grade beam and slope away from the building. Interior finish floor should come up no less than 2 inches (5 cm) below the top of the grade beam to prevent moisture spilled on the floor from broken pipes or overflowing bathtubs and so on from seeping into walls.

The rubble trench foundation and French drain do a spectacular job of removing moisture from around a house. The crushed rock or pumice, for example, permits water to drain down into the trench where it is drawn away by the French drain. Because it removes moisture so effectively, this foundation system prevents moisture from wicking up (moving by capillary action) into the overlying wall system, protecting earthen and straw bale walls and plaster from moisture damage.

CONCRETE FOUNDATIONS. Some owner-builders and professional builders choose more conventional poured concrete foundations for straw bale or natural homes. Concrete foundations for conventional homes typically consist of two parts, a footing and a stem wall. The footing distributes the weight of the structure over a wider area, much as your feet do. The stem wall transmits the load from the walls, floors, roof, and so on to the footing, as your legs do. Because straw bale homes require a wide foundation (stem) wall, many builders dispense with the footing. They simply excavate a trench for the foundation, then erect forms for a wide stem wall that will support the width of the straw bale walls and the weight of the roof.

Another popular option is the monolithic slab (also known as a slab-on-grade foundation) which combines a concrete slab (subfloor) and stem wall in one pour.

If you elect to build a concrete foundation, care should be taken to prevent water in the ground from seeping into the wall and wicking up into the bales or earthen walls. Both conventional stem wall and slab-on-grade foundations benefit from the installation of an external perimeter French drain. Also of benefit is the use of a good non-toxic foundation sealer. Although there are several products available for sealing foundations and grade beams, we prefer a product known as Dynoseal (manufactured by AFM) over standard petroleum-based foundation sealers sold in major building outlets and hardware stores. AFM Dynoseal not only acts as a moisture barrier, it serves as a barrier to radon gas. Moreover, it does not release toxic substances, as do commonly used products, which are known to have long-term health impacts. Dynoseal is therefore not only effective, it is far safer than the standard sealants on the market. (When Dynoseal is used in areas exposed to sunlight, be sure to apply a coat of Dynoflex over it to protect the Dynoseal from exposure to ultraviolet radiation. Dynoseal is black/brown in color; Dynoflex is white and can be tinted.)

Whatever you do, don't lay plastic on the top of a stem wall or grade beam or wrap bales with plastic. When bales rest on plastic, water can condense on the plastic between it and the straw bales. This causes straw to mold and rot, eventually leading to failure. Never use roofing felt as a moisture barrier, either. This product is not a good water barrier and it contains a number of hazardous chemicals that release gas into your home for years. AFM Dynoseal, in contrast, creates a well-sealed stemwall or grade beam.

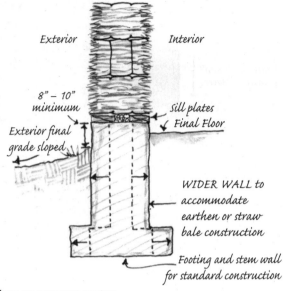

Exterior   Interior

8" – 10" minimum

Exterior final grade sloped

Sill plates
Final Floor

WIDER WALL to accommodate earthen or straw bale construction

Footing and stem wall for standard construction

3-5 (a): Typical concrete foundations consist of a stem wall and footing, and can be used for straw bale walls. Note the sloped grading to promote water drainage.

3-5 (b): The monolithic slab is also popular among straw bale builders. Note edge of slab is widened to support straw bale wall.

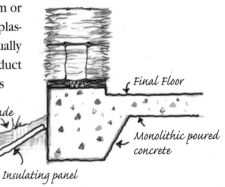

Final Floor

Final grade

Monolithic poured concrete

Insulating panel

3-6: Locally available stone creates a strong and beautiful foundation for straw bale and earthen walls. Because it can often come from the building site, it is considered a low embodied energy material. Stone does not wick moisture into a wall the way a concrete foundation does.

Rather than setting straw bales directly on a sealed grade beam or stem wall, however, we recommend installing another measure of protection — a sill or bottom plate. You can build a bottom plate out of reclaimed or sustainably harvested redwood, cedar, or fir — for example, 2 x 4s or 2 x 6s — treated with a non-toxic sealant. This wooden structure is bolted in place on top of the sealed grade beam or stem wall. A layer of plastic sill seal (pink or gray plastic) is sandwiched between the foundation and the sill plate to prevent moisture from wicking into the wood. The bottom plate serves as a moisture break that adds another level of protection.

STONE AND WOOD FOUNDATIONS. Traditional dry-stacked or mortared stone makes an excellent foundation for straw bale and earthen walls. Whether designed and built for a deep frost-proof foundation or as a grade beam over a rubble trench, this time-honored system can provide a strong, stable, water-resistant foundation. Not only do they look good, stone foundations eliminate or reduce the use of high embodied energy cement.

Reclaimed or sustainably harvested redwood, cedar, and cypress can also be used for building several types of post-and-beam foundation systems. Charring the wood before burying it or coating it with a healthy, natural wood treatment can ensure a longer-lasting foundation. Never use railroad ties; they contain numerous toxic chemicals that can release gases into a home. If you are required to use pressure-treated wood, use arsenic- and chromium-free ACQ treated lumber and seal it well to prevent releasing of gases and leaching of chemicals.

3-7: Earthbag grade beam on stone and rubble trench foundation. Level is being used here to be certain top of foundation is level, which is vital for placement of the first course of staw bales.

EARTHBAG FOUNDATIONS. Earthbags are often used to make reliable and inexpensive foundations for small natural homes. Earthbags are often used to make grade beams for rubble trench foundations.

As noted in Chapter 1, Earthbag builders typically use 50-pound (22 kilogram) polypropylene grain sacks or burlap bags. The bags are filled with a slightly moist soil containing approximately 30% clay and 70% sand and aggregate, then tamped. When the soil mix dries, the bags become hard as sandstone. When building foundations, many natural builders have begun to install a couple of courses of bags filled with crushed rock or compacted gravel to retard water movement into the foundation.

Protecting the Transition Zone

Many builders have demonstrated enormous ingenuity when it comes to foundations, using a variety of creative systems that ensure the required stability and moisture protection. No matter what type of foundation system you select, be sure that the top of the foundation is sufficiently

above final grade. Be sure to seal the foundation, too, and grade the dirt around the structure so that it slopes away from the foundation. Install a French drain at the perimeter of the building. Combined, these measures help to eliminate moisture problems.

It is also important to prevent plaster from coming into contact with moist or wet ground. Some builders run plaster down their walls onto their foundations, which then becomes wet. Moisture also creeps up from the ground, attacking the plaster at the base of the wall. (See photo 2-10). To prevent this, you may want to install flashing along the top of the foundation or grade beam, to indicate where the plaster should stop — or you can also simply mark a line on the foundation wall sufficiently above grade. Apply plasters to the wall and take them down to the edge of the flashing or to your line. Some builders apply sealer to the first 18 – 36 inches (45 – 90 cm) of a plastered wall to protect them from moisture. However, many sealers contain very toxic chemicals that can affect your health. They also create a non-breathable surface. For a successful earthen home, you need a water-resistant yet breathable plaster. Many natural additives will achieve this goal, as will be detailed in Chapter 4.

3-8: Backsplash from roof runoff can cause serious water damage to walls. An inadequate foundation as seen here (for example, when the bale wall is not raised sufficiently off the ground) may result in water wicking up into the wall, damaging both straw and plaster.

3 good ways to establish the bottom edge of the plaster

Plaster

WRONG!

Plaster stop

Flashing

A line on the foundation

3-9 (a) (b) (c): Exterior plaster should not extend to grade. Stop at (a) plaster stop (b) flashing or (c) establish a line sufficiently above grade.

Stone facings applied to the foundation and lower wall represent yet another important means of protecting lower wall surfaces. Stones are applied in several ways to the foundation surface and to the first 18 - 36 inches (45 - 90 cm) of the wall: for example, stone can be can mortared on the external surface of the foundation, then continued up the face of the wall, using anchor pins to secure them. Or you can start with a wide base of stone (mortared or dry stacked) and build those up to cover foundation and 18 - 36 inches (45 - 90 cm) onto the wall. Rock facings prevent backsplash from eroding the earthen plaster on the lower wall surfaces. Backsplash is water that falls on the ground around your home either from rain or from water pouring off the roof. It hits the ground and splashes onto the lower reaches of walls. To be

Avoid "Backsplash" with this option

• *Protective stone facing*
• *Pea gravel at drip line*

*6" – 8" deep trench filled with a pea gravel "sponge"
or French Drain as in illustration 3-2*

3-10 (a) (b): Rock facings can be used to prevent backsplash from damaging plaster along lower reaches of the wall. Pea gravel-filled trench at the dripline reduces backsplash.

on the safe side, we recommend installing a protective stone facing even when gutters are used. This will protect walls from rains if gutters become clogged and overflow. (We recommend that you install gutters but don't count on them to work all of the time!)

Roofs and Walls

When designing the hat or roof of a home, it is the details of roof design (style, overhangs, intersections, abutments, and gutters) that determine how well a roof will protect walls.

Before you can design a successful roof, you need to have a good understanding of local climate conditions, as noted earlier, so that you can take into account how prevailing weather conditions will interact with the building. When a north/west storm comes in and the rain hits your roof, where will it go? How will it respond? How will it affect your walls and foundation?

3-11: Unprotected earthen plaster walls (in this case there was no overhang) can be severely damaged by driving rains, as shown here. If not periodically refinished or repaired, damage can become serious.

CEDAR ROSE GUELBERTH

ROOF DESIGN CONSIDERATIONS. It is important to look at the ways a roof design interacts with and sheds water. Each section of a roof collects water and sends it to another location, usually the ground. The smaller the roof section, the less water is shed to a given area. Larger roof sections obviously shed larger quantities of water.

Remember: a roof's principle function is to protect a building, including walls and wall finishes. A roof should be designed so that it can shed water to designated areas or can evenly distribute it. Roof intersections funnel water to specific areas, and large, flat areas distribute water more evenly, depending on the direction of a storm.

Shedding water away from walls is paramount to the success of a plaster or stucco. Designing a roof with consideration given to climate conditions can achieve this protection. The pitch (the angle or slope) determines how each section of a roof drains and the speed at which it drains; abutments between roof segments; abutments between walls and roof sections; and roof styles (hip, gable, flat, Dutch hip, etc.), all need to be evaluated in relation to how they interact with weather at your site. By educating yourself to your specific climatic concerns and designing roof details to meet and solve those concerns, you can protect wall plaster and achieve a successful long-lasting structure.

Earthen plasters are by nature designed to handle moisture; however, regular driving rains will eventually cause wear. A well-designed roof system can minimize this type of wear.

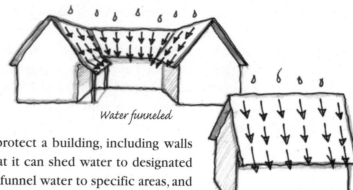

Water funneled

Water evenly distributed

3-12 (a) (b): Roofs protect walls, but the design can have profound effects on where water goes. In the top drawing, water from a large area is funneled to a relatively small area, resulting in potential accumulation around the house. In the lower drawing, water is distributed over a relatively larger area, reducing the potential for buildup and potential problems with the foundation and walls.

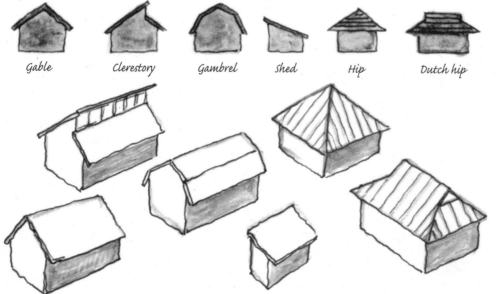

Gable   Clerestory   Gambrel   Shed   Hip   Dutch hip

3-13: Common roof styles. Each type of roof sheds water differently. Be aware of the direction storms come from, how they interact with the roof design, how water will be distributed, and where it will go.

3-14: Attention to roof and wall design is essential. In the top drawing, the roof design sheds water nicely. In the bottom drawing, the roof abuts an earthen plastered wall, where rainwater and snowmelt can damage plaster and seep moisture into adjoining walls.

A good detail – Roof NOT adjoining strawbale wall

YES

A gable roof with a shed roof under, sheds water away from wall

OK

Bad construction detail – Roof adjoining strawbale wall

NO

3-15: Large, unprotected walls as shown here are vulnerable to wind and water, which can ruin plaster and damage the underlying wall system.

DAN CHIRAS

OVERHANGS. Overhangs (eaves) are an especially important aspect of the roof design. They protect walls from rain and are vital when earthen plasters are being considered. In fact, a good overhang prevents wear in all but the most severe storms.

For many homes, a 24 inch (60 cm) overhang is sufficient. Wider eaves may be needed in rainier or more severe climates. Covered porches also provide excellent protection for walls. In some regions, you may want to vary the overhangs on different sides of a building, depending on where the storms come from.

Along the south side of a home, be aware of passive solar heating potential. Sunlight can be a valuable source of heat. If a home is designed carefully and you live in an area with ample sunshine, passive solar gain can provide all of your heat, freeing you from costly fuel bills and exonerating you from the damage fossil fuels are creating. In a passive solar home, however, too much overhang on the south side of a home can greatly reduce sunlight penetration, especially during the fall and spring months. There are ways, however, to maximize solar gain while still protecting earthen plasters.

ABUTMENTS. Not all houses are simple rectangles with a single roof. Many homes are designed with complex roof lines that intersect with second stories and dormers or other structural features. In the such instances, be sure that roof lines do not abut straw bale or natural walls: snow accumulating at this interface or rainwater running along the junction of the wall and roof line will erode the earthen plaster. No matter how well you install flashing, these areas will require periodic maintenance and may permit moisture to penetrate the straw bale walls. Moisture and straw don't mix. That's a prescription for disaster!

DESIGNING AND PROTECTING WALLS. To create a successful structure, the design of a wall system in relationship to the roof is important. We will discuss the details of straw bale wall systems momentarily. Natural earthen walls were described briefly in Chapter 1 and in more detail in Chapter 10.

One of the most effective ways to ensure successful earthen finishes is to create smaller, protected wall surfaces. Eliminate large open wall expanses that expose walls to severe weather. Tall, two-story wall spans leave a wall extremely vulnerable to the weather.

Be aware of roof designs that can make walls vulnerable, too. Gable roofs, for instance, are relatively simple, easy-to-build, and inexpensive; however, the gable ends of a building are generally less well protected and thus more vulnerable to the weather. Design the gable ends of homes and other buildings to reduce the exposed surface — that is, break up the expansive surfaces into smaller protected wall sections. You can protect the exposed upper portions of walls by applying wood siding, adding a porch roof, or creating a latticework fascia.

Houses can also be sited to take advantage of the protection of trees and bushes. Trees and bushes can be planted around the house to act as a buffer from driving rain. Be sure, however, that they are planted well away from your structure: moisture released by trees and shrubs can affect the success of your plasters. In addition, trees, shrubs, or other plants growing close to walls can trap moisture against the wall, which can eventually lead to serious problems. Although you may be tempted to grow ivy on a wall to protect it, don't do it! Ivy traps moisture and attaches securely into the wall surface, which can seriously damage plaster.

If a home is designed well, water running from the roof should reach the ground without running down its walls. As noted earlier, water falling to the ground can create backsplash onto the plaster on the lower reaches of the wall, eroding the plaster and eventually damaging the underlying wall system. When designing a foundation and walls, it is critical to include elements that prevent the occurrence of backsplash. There are relatively simple solutions to this problem, as described earlier in the discussion of foundations. You can end the plaster well above final grade and protect the lower section of the wall with a protective stone facing, and create a good perimeter drainage system around the house.

Another protective mechanism is a perimeter zone of rock — created by filling a shallow trench with stone at the "drip edge" line. To install a perimeter drainage system, first determine where the drip line occurs. Then dig a 6 - 12-inch (15 - 30 cm) wide and 6 - 12 inch (15 - 30 cm) deep trench around the perimeter of the home, and fill it with gravel or crushed rock. (Pea gravel works well, see illus. 3-6) The perimeter rock zone creates an absorbent surface, so that, rather than backsplashing, water is absorbed by the rock "sponge." In high-moisture areas where the ground becomes saturated quickly, you will need to dig a deeper and wider trench, 12 - 18 inches (30 - 45 cm) wide and deep. Slope the bottom of the trench "to daylight," install a French drain and cover it with gravel. (See illus. 3-2.) This system not only prevents backsplashing, it drains water away from the perimeter of the foundation, preventing water from pooling around a home.

3-16: Gable ends of homes leave walls vulnerable to the effects of weather. By siding the upper portion of an exposed gable end, increasing the width of the fascia, or breaking up the roof line, builders can protect walls from rain and other erosive forces.

You can also prevent backsplash by installing gutters and downspouts, creating a direct route for water to move from the roof to the ground, completely bypassing the walls. Although recommended for straw bale and earthen structures, gutters and downspouts should be considered a second line of defense (gutters can fail). Roofs, walls, and foundation systems should be designed and crafted to accommodate weather, whether you have gutters or not.

By transporting water from the roof to the ground, gutters prevent water from striking the wall on its way down and/or splashing back up after it hits the ground. To be effective, gutters must empty into downspouts which carry the water down to the ground and away from the building. If you're catching roof water and storing it in a cistern, water will be transported from the downspouts to your storage tanks via a series of pipes. These pipes prevent water from accumulating around the foundation. If you're not collecting water, downspout extenders can be installed to carry water well away from a building.

Gutters provide a measure of protection, but don't be too dependent on them. They can become clogged with leaves and other debris or may receive too much water at once during heavy rainstorms and may overflow. That's why it is important to design a home to handle rain water without gutters, then include them as a second line of defense. They can also play an important role if, after the fact, you discover areas where water runoff from the roof exceeds expectations. Be sure to design a gutter system to accommodate the maximum volume of water and remember to clean them regularly.

In summary, true protection from rain requires adequate overhangs, perimeter drain zones, rock facing along the lower portion wall and foundation, and gutters. These elements are effective in handling a lot of different types of rain, though driving rains are more problematic. They require designing smaller wall areas, stone-facing on the lower reaches of walls, and attention to detail around windows and doors, our next subject.

PROTECTING WINDOWS AND DOORS. If a roof is well designed, windows and doors mounted in walls should have minimal exposure to rain or snow, except in driving rain or drifting snow. Several precautions can be taken to protect your plaster (and ultimately your wall system) around windows and doors.

3-17 (a) (b): Proper window design and construction is vital. Minimal exposure of the exterior sill reduces water capture, and an angled drip sill of tile or stone will drain water away from the window and underlying wall to prevent damage.

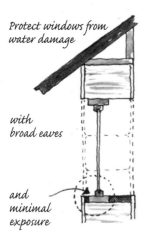

Protect windows from water damage

with broad eaves

and minimal exposure

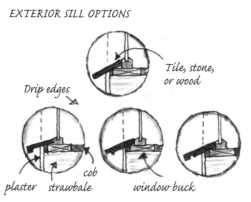

EXTERIOR SILL OPTIONS

Tile, stone, or wood

Drip edges

plaster strawbale cob

window buck

Windows and doors should be aligned so they are flush with the exterior plaster or slightly inset, so water won't accumulate along the frame or drip down the wall. (Placing a window in this manner results in deep window sills on the interior of a home, which many people find attractive.)

Take extra care with window details. Many builders, for instance, use tar paper, visquine or metal lath to line straw bales, window sills, and the sides of their window openings and window bucks before plastering. However, these techniques can cause serious problems for a number of reasons. First, non-breathable materials such as visquine can trap moisture against the

bales, causing mold to form and straw to rot. Second, plasters don't adhere at all to tar paper or other nonporous surfaces. Although chicken wire or metal lath may be applied to the waterproof lining to create a surface for plaster to attach to, they provide only superficial and temporary anchorage. That is to say, mesh (chicken wire or metal lath) only provides a superficial surface with "tooth" for the plaster to key into. A straw bale provides a far better surface to key into, if prepared correctly. In addition to these problems, air pockets may exist beneath the metal lath or chicken wire. Over time, these areas are much more likely to deteriorate, creating serious problems. In contrast, natural clays present in the coblike packing mixture that is keyed into the bales provides a superior base. Combined with clay present in the plaster itself, it provides an excellent protective coating for natural walls.

When building a home, it is better to pack a mixture of straw and clay-dirt around window and door bucks (rough frames), from which you can build a stable coblike

earthen surface that provides a strong, thick, stable base for plaster. Using these materials also allows for beautiful curves and lines around windows and doors, giving you the freedom to shape these areas as you would like them.

If gaps between window or door buck are large, screws can be driven into the wood to form a secure anchorage for the earthen filler and the plasters you will apply to it. (Heads of the screws need to protrude about 1/4 - 1/2 inch or 60 - 120 mm). This provides a strong anchor for infill and plaster.

3-18: Metal lath around windows and doors can cause serious problems with plaster, as explained in the text. We suggest other techniques for preparing walls for plaster.

3-19: Packing a mixture of straw and clay around window and door frames then building out with a coblike mixture, as shown here, creates a strong, stable base for plasters — far superior to lath and other materials.

CEDAR ROSE GUELBERTH

3-20: Screws protruding from framing materials help anchor plaster to wood surfaces. Additional "tooth" is provided by an adhesion coat, shown here and described on page 61.

CEDAR ROSE GUELBERTH

CEDAR ROSE GUELBERTH

3-21: This wood eyebrow and sill protects the window and prevents water from dripping down the wall.

For further protection against moisture for recessed windows, you should apply a sloped stone, tile, or wood sill — or metal flashing — along the base of the window to shed water away from the window and wall, as seen in figures 3-17 and 3-21. Be sure to extend the material far enough to create a "drip edge" to prevent water from running down the wall. (Water running down plaster from windows can cause a great deal of damage.)

Another means of protecting windows is to build "eyebrows" over them. Eyebrows can be sculpted with earthen materials, such as a cob sculpting mix, or made from carved wood, metal flashing, or stone — combining functionality and creativity.

Eyebrows serve as a diversion structure to shed water away from a building, as in a mini awning, or simply to divert water away from a window and window frame, preventing it from accumulating on the window ledge or working its way between the plaster and the window frame. However, it is important to note that cob eyebrows only work in areas that don't experience "pouring rains." Heavy rainfall can create erosion along the walls, resulting in a lot of maintenance.

PROTECTING INTERIOR WALLS. Most of this discussion has focused on exterior walls, since they are more vulnerable to the elements, but interior wall surfaces also need protection. When building an earthen or straw bale structure, walls should be raised off the final floor at least 2 inches (5 cm) by the grade beam, as shown in illustration 3-3. That way, if a pipe breaks or a 100-gallon aquarium leaks or a bathtub overflows, water on the floor that comes in contact with the plaster won't seep into the wall.

Water vapor in the air on the inside of a house can also be a problem which deserves special design consideration. In attempting to create energy-efficient homes, many designers and builders have made "improvements" that cut down on air infiltration and rightfully so. By sealing homes more tightly, they conserve energy. However, such measures can result in a number of problems in both conventional and earthen-plastered homes. While earthen plasters can "process" the flow of ambient moisture well, it is possible, as with any other finish, that they will develop mold and mildew in areas that experience constant high humidity — for example, poorly ventilated bathrooms. To prevent this problem, homes must be designed to provide air circulation and ventilation.

Good ventilation reduces moisture buildup and protects against mold and mildew — and their subsequent health effects. It helps to ensure good indoor air quality. Open floor plans, easy-to-operate and well-planned windows for cross ventilation, or automated heat recovery ventilation (often called air-to-air heat exchangers) controlled by a humidistat are all recommended.

Anatomy of a Straw Bale Wall

The success of plasters hinges on many factors, such as wall design, wall construction, and how the walls are prepared for plastering. These factors also affect the ease with which plaster goes onto the wall and the time it takes to plaster a wall. In order to create a good surface for a natural plaster finish, the builder using straw bale needs to pay particular attention to the factors that will affect the plastering itself.

Straw bale walls come in two basic varieties: load bearing and nonload-bearing. A load-bearing wall is, as the name implies, built in such a way that the straw bales themselves support the roof. That's right, the straw bales bear the load placed upon them — and they do so admirably well. Much to the surprise of structural engineers, a load-bearing straw bale wall supports nearly seven times more weight than a two-by-four-stud wall!

Good quality bales are laid on a suitable foundation stacked in a running bond — that is, an overlapping pattern like stones or bricks in a wall. This locks the bales together and increases the rigidity of the wall. (Continuous vertical joints in straw bale walls should be avoided!)

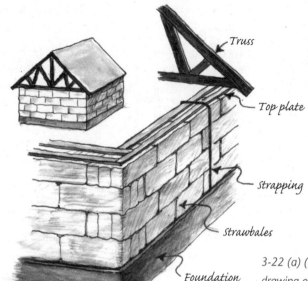

3-22 (a) (b): In the top drawing of a load-bearing straw bale wall, the weight of the roof rests directly on a top plate which rests directly on the bales.

In the bottom drawing of a nonload-bearing wall, the weight of the roof is supported by timber frame posts and beams. Bales are laid on a foundation and do not support the roof.

During wall construction, further rigidity (which increases resistance to forces such as wind) is ensured by pinning the wall. Pins made of steel rebar or bamboo may be driven into the bales, a process called internal pinning, or may be attached to the inner and outer surface of the straw bale walls, a process known as *external pinning*. Unfortunately, unless external pins are notched and tightly cinched to the wall, they're often difficult to plaster over. Plastering walls that are internally pinned is far easier.

EXTERNAL PINNING

INTERNAL PINNING

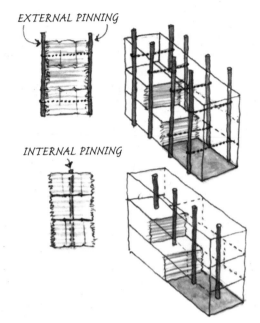

3-23 (a) (b): Internal pins help stabilize a straw bale wall. External pins provide structural stability, but present more of a challenge when it comes to plastering a wall.

In load-bearing straw bale walls, a top plate is secured to the bale wall. The top plate, also known as the roof-bearing assembly, serves several important functions. First of all, it provides a means of attaching the roof to the walls of your home. It also distributes the weight of the roof evenly on the walls, preventing buckling of the walls due to uneven loading. Finally, the top plate (and the roof) helps to make a straw bale home more solid. That is, it holds the structure together.

To anchor the roof to the walls, however, the top plate in a load-bearing wall system should be attached to the foundation. One of the most popular systems anchors the top plate to the foundation via polyester straps, heavy wire, or aircraft cable, which provide a way to cinch down the top plate, applying pressure onto the bales — that is, compressing them. (The importance of this will be evident shortly.)

At one time, many straw bale builders used a product called all-thread to provide anchorage. All-thread is a fully threaded rod that runs from the foundation to the top plate. It is located internally — that is, inside the straw bale walls. All-thread has fallen into disfavor, because it is time-consuming and more difficult than other techniques. (Further details are found in the straw bale building books listed in our Resource Guide.)

The second type of straw bale wall is known as a nonload-bearing wall. It does not support the roof system. Rather, the roof is supported by a structural frame, typically a traditional timberframe, post-and-beam, modified post-and-beam, or stud frame structure. Your options for frame materials are many. Some builders use natural logs collected on or near the site for posts and beams. Others use dimensional lumber milled from locally harvested trees or local lumber suppliers. Still others used certified ecologically harvested or reclaimed materials. In some instances, straw bale builders have even used concrete blocks, but because concrete has a higher embodied energy than wood, we urge readers to consider other options. Some builders have even used steel studs in straw bale construction. Steel production, however, has a high environmental impact and for those concerned with magnetic fields it may not be the best choice. It should be clear, however, that your choices are many and most of them are amenable to earthen plasters. Some of them can yield rather stunning results.

There are several ways to construct nonload-bearing straw bale walls: infill, wrap, or partial wrap. In the infill method, bales are stacked between the framing members.

Another, and often easier system, is the straw bale wrap. When "wrapping" bales around a frame, the bales are laid down in a running bond pattern on either the interior or the exterior of the frame. If the bales are on the exterior, the post-and-beam frame will be visible inside of the home. In the wrap technique, the straw bales form a continuous thermal envelope, eliminating heat loss through seams between bales and framing members (known as thermal breaks). Another system is a partial wrap, which is similar to the system just described, but the frame is either completely or partially embedded into the wall.

In all of these systems, earthen plaster is applied to the straw bale wall, creating a living protective skin that allows air and moisture vapor to pass through it, slowly and gently maintaining a comfortable, healthy interior environment. As shown in the accompanying illustration, the plaster may completely cover the framing members or may be applied up to them.

One thing that is worth noting is that building department officials often prefer the non-load bearing walls over the load-bearing wall, even though tests show that the latter are sturdy enough to support considerable weight and many load-bearing structures have been built. Why? For one thing, post-and-beam structures are familiar to them: they're used in a variety of applications such as barns, sheds, and houses. Code officials feel comfortable with post-and-beam construction in straw bale and other natural buildings because they understand the physics of the structure. Moreover, they can use charts and tables to determine the size of posts and beams for different roof loads.

"INSET" Framing

FRAMING OPTIONS

"INFILL" between framing

"EXTERNAL WRAP" or "INTERNAL WRAP"

3-23 (a) (b) (c): Three ways to build nonload-bearing walls. (a) In the infill method, bales are placed between the posts. (b) In the wrap technique, bales are laid outside or inside against the post-and-beam structure. (c) In the partial wrap, the bales are notched and laid in such a way that they encompass the posts.

Plaster up to posts leaving them exposed;

or plaster over posts, covering them.

3-24 (a) (b): Plaster details in post-and-beam construction. (a) Plaster up to post, leaving post exposed. (b) Plaster carried over framing.

Remember to apply an adhesion coat to wood before plastering.

Code officials may also be more amenable to the use of earthen plasters in a nonload-bearing wall design because earthen plasters are not considered essential to the structural integrity of the wall. In such instances, the frame carries all of the weight. In contrast, in some load-bearing straw bale designs, cement stucco is considered an important component of the structural system. (Our recommendation is that you not depend on the inch or two of stucco on interior and exterior walls to hold up your building. Your building — no matter if it designed as a load-bearing or nonload-bearing straw bale structure — should be designed in such a way as to provide its own structural support, allowing the plaster or stucco to perform its task of protection for your walls.)

With this overview in mind, we turn our attention to specific construction details that enhance the success of an earthen plaster.

Choose Bales Carefully

For successful plastering, straw bale and earthen walls must be constructed well. Straw bale walls also need to be prepared for plastering — that is, they must be "shaved" and shaped. We'll begin this discussion with a brief look at bale orientation.

As you may know, straw bales come in two- and three-string varieties. The dimensions of straw bales vary, depending on the baler and the settings selected by the farmer. Generally, two-string bales are 17 – 18 inches (43 – 35 cm) wide while three-string bales are 23 – 24 inches (58 – 60 cm) wide and considerably heavier. Both types are commonly used for construction. While the two-string bales may be easier to lift and move, the three-string bales provide more insulation.

The most important thing to keep in mind is the quality of the bale and the straw. You want bales that are highly compacted (very dense) and solid. Strings should be tight and bales should not bend when lifted by the strings. Buy bales tied with strong polypropylene baling twine. Be sure the straw is of good quality — that is, the straw bales are dry, straw is a light bright color (avoid bales showing signs of discoloration) — and that the bales contain long shafts of straw. (Short strands of straw may be great for plaster mixes but for house construction, bales should be made from long straw.)

Building Bale Walls to Provide Maximum Benefit

Most people build the walls of their homes by laying their bales flat, that is, with the strings up, as opposed to laying them on their sides — which is fortunate, as it provides a stronger base with more "tooth" for a plaster to key into. When a bale is laid flat, the exposed surfaces on each side of the wall — that is, the surfaces to which the plaster will be applied — are rough. They consist of the ends of many shafts of straw that stick out more or less horizontally. Because the exposed surfaces are rough, the bales accept plaster well. Thus, when bales are laid flat, the first coat of plasters key into them very well. This, in turn, provides a solid base for subsequent layers of plasters.

When bales are laid on edge, earthen plaster goes on but not as well. The exposed surfaces to which the plaster will be applied consist of many tightly packed, parallel strands of straw. Straw itself has a waxy, nearly glossy coat. As a result, the flat surface is smoother and is more difficult to key in a plaster. Don't get us wrong: earthen plasters will adhere to the surface, just not as well. (You are counting on the packing of straw/clay in the seams between bales, as well as a good high-pressure clay slip application, to key in your plasters securely.)

When laying bales flat with the strings facing up, you will notice that one side of the bale is rougher than the other. This is due to the way bales are produced. For best results, always place bales in a wall so that the rougher sides face the exterior surface of the building. This provides a "toothier" surface that holds the plaster better. Exterior plaster can use this extra advantage. Although interior plaster needs to key into the wall, it is not as critical for interior plasters, which are exposed to much milder conditions. Besides being easier to plaster and producing a better bond between plaster and straw bales, flat-bale orientation provides a higher level of insulation. Whether you choose two-string or three-string bales, our advice is to lay them flat.

Prepping Walls for Plaster

When bales are being laid, be sure that the walls are plumb and solidly built no matter whether the walls are straight or curved. This is especially important if you're sponsoring a workshop where overzealous fledgling straw balers can become a bit careless. Use a large wooden mallet, a sledge hammer, or a tamper to straighten the bales.

Bales laid on their sides with strings facing out do not accept plaster as well as...

Bales laid flat with strings facing up.

3-25 (a) (b): Bale orientation.

3-26 (a) (b): Worker makes sure bales are level (left). A large wooden mallet is used to adjust bales for an even wall surface (right).

CEDAR ROSE GUELBERTH   CEDAR ROSE GUELBERTH

When building, you'll want to be sure that walls remain as plumb as possible. A carpenter's level and a wooden mallet are vital to this task.

Work slowly and resist the temptation to build your walls in a day, as often happens when a group of workers catches the dreaded bale frenzy. Be aware of protruding bales at corners and around door and window openings. Building slowly with a watchful eye will help you avert those little problems that often snowball over time.

SHAVING AND SHAPING WALLS. Stacking bales evenly — that is, ensuring walls are plumb — produces a solid, even base for plasters. However, even walls still need to be trimmed or "shaved" before plastering. Trimming the exposed surface gets rid of loose ends of straw, evens out bumps, and helps to shape the walls (contributing to its final appearance), all of which will make plastering much easier. Remember that irregularities in a wall surface will stand out as soon as the first coat of plaster goes on. The better the job you do trimming the walls, the better the final product — and the easier the plastering will be, which saves time!

Many builders use electric weed trimmers (known colloquially as weed eaters or weed whackers) to trim bale walls. Detail shaping can also be performed with an electric chain saw, which is easier to use than a gas-powered chainsaw, as well as being somewhat quieter and posing less of a fire hazard. Electric chains saws and weed eaters can be tilted and worked at odd angles without spilling gas or having their engines cut out.

To trim a wall, go over the surface with the weed eater, slowly and methodically, evening out the surface, and removing loose straw. Be sure to wear a dust mask

3-27 (Left): Weed eaters and electric chains saws are used to trim loose straw and shape bale walls in preparation for plaster.

3-28 (Right): Keep walls plumb and even. Watch out for protruding bales while building walls, especially at corners and around windows and doors, as shown here.

CEDAR ROSE GUELBERTH

CEDAR ROSE GUELBERTH

STUFFING AND PACKING WALLS. Before shaping the walls, many straw bale home builders pack the joints in the walls — the gaps between bales — with loose straw. Although this is important, we recommend doing this later, after you apply the slip coat. (The slip coat is a thin layer of clay-rich dirt that's applied wet to the surface of straw bale walls prior to plastering, discussed in Chapter 6.) We also recommend stuffing deep gaps first with dry straw, then finishing the job with a wet straw/clay-dirt mix to create a more solid, secure surface. Packing only with loose straw leaves unsecured fibers that create an unstable base for plasters to key into, and also creates air pockets — which together result in unstable surfaces or weak points in the plaster. Plaster applied to these areas can deteriorate, and may even fall or peel off the wall. The system we'll present in Chapter 6, we believe, results in a more stable surface with better long-term performance.

INSTALLING ELECTRICAL WIRING AND PLUMBING. Once walls have been trimmed and shaped, it is time to install electrical wiring, plumbing, external pinning (if used) and strapping (for load bearing). To install electrical wiring, be sure to nestle the wiring into the bales — running it along seams between bales is easiest. Wherever possible, eliminate wire running along the surface of bales, especially bundles of wires. If necessary, recess wiring by notching bales. In some cases, building codes require that electrical wiring be installed in plastic conduit, which makes wiring even bulkier and more difficult to plaster over, so be sure to embed it into the surface of the bales.

Be certain that electrical boxes for lights, outlets, and switches are securely attached to walls as well. Boxes are generally mounted in notches carved out of the straw bale walls and secured by stakes driven into the bales or are attached to framing members. When installing boxes, be sure to keep in mind the depth of the plaster — that is, place the boxes so that the final coat will be flush with the outer surface of the box, including the attached plaster ring. (A plaster ring is a device that fits onto the electrical box, which is used to provide proper depth for final plaster. Different sizes are available to mount onto your boxes to accommodate the finish coat.)

Most builders refrain from installing plumbing pipes in straw bale walls for fear of leaks and condensation which can damage bales. If you need to, remember to "double case" all water pipes (enclose copper pipes in a plastic pipe) and notch the pipes into the walls so you can plaster over them. Gas lines and vent pipes often end up along bale walls and need to be installed tightly against the bales or, preferably, notched into the bale surface. Otherwise, you may need to build out the wall with a cob-like mix (a mix of clay-dirt, sand, and straw) to sculpt a creative covering for the pipes or build a chase to house them.

PINNING WALLS. For most nonload-bearing straw bale construction, builders pin their bales internally. This is the best system for plastering because the pins are "buried" in the middle of the wall system: you don't need to plaster over them. However, while internal pinning provides effective stability for nonload-bearing walls (post-and-beam construction), many natural builders believe that

CEDAR ROSE GUELBERTH

CEDAR ROSE GUELBERTH

CEDAR ROSE GUELBERTH

3-29: Here wiring has been inserted along bale seams. Be sure that all exposed wire (and conduit, where required by local building codes) is embedded in the bale surface to avoid unsightly lines or bumps in the finished plaster.

3-30: Mounting electrical boxes at the correct depth and installing the correct size plaster ring (the plate attached to the top of the box) are critical to achieving an aesthetically appealing wall. (Note that the plate that the switch is being screwed into is the plaster ring).

3-31: Building out a wall to cover pipes avoids an eyesore and can be quite intriguing as in this sculpted tree covering pipes.

external pinning provides more stability for load-bearing walls. Internal pinning is done while the bale walls are going up. External pinning is done after the bale walls are up and after the surfaces have been trimmed and shaped.

When building load-bearing straw bale structures, many people use strapping or cables to attach the top plate (and the attached roof) to the foundation. External pins are used to stabilize the wall. When using strapping and exterior pins, there are simple things you can do to make the plaster job easier.

"Ideally," veteran straw bale builder Steve Kemble advises, "the top plate should be slightly smaller than the width of the straw bale wall. That way the straps will be more likely to run flush against the inner and outer wall surface of the straw bale walls. If the straps or cables run off the wall, you can sew them into the wall to draw them closer."

You can notch the top edge of the straw bales along the top plate, so external pins and straps fit more snugly against the wall, and notch the wall itself with an electric chain saw so the pins can be nestled tightly against the bales. Pins should also be cinched tightly against the surface of the bale by sewing the pins together — that is, cinching one to the other through the bale wall using twine. Cinching and nestling pins and cables into the wall will make plastering easier. (Note that notching and cinching techniques only work when bales are laid flat).

When building a load-bearing wall, it is important to let the walls fully settle before applying earthen plaster, except perhaps the very first layer, the scratch coat; otherwise, the plaster will very likely crack as the bales settle over time. Settling can be achieved in one of two ways. If you've built walls with straps or cables, you can precompress them by uniformly cinching down the connectors. (See *Build it With Bales* for more details on this procedure.) Installing a roof system also provides compression and

3-32: In this illustration, you can see many of the important details for building a wall so that it will be easy to plaster. Note that the strapping is cinched to the wall and the external pins are placed in notches and then secured tightly to the walls so plaster goes on easily without creating air pockets.

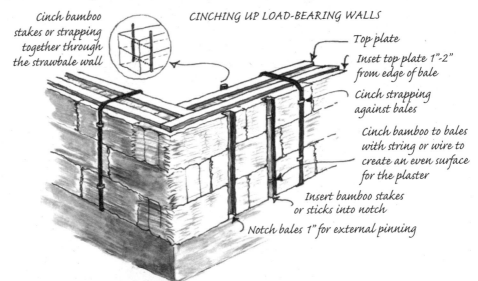

Cinch bamboo stakes or strapping together through the strawbale wall

CINCHING UP LOAD-BEARING WALLS

Top plate

Inset top plate 1"-2" from edge of bale

Cinch strapping against bales

Cinch bamboo to bales with string or wire to create an even surface for the plaster

Insert bamboo stakes or sticks into notch

Notch bales 1" for external pinning

allows you to uniformly tighten and cinch down the connectors over a period of time. The combination of the weight of the roof and the cinching of cables (straps) does a good job of compressing the walls. In some cases, walls will compress minimally; other times, walls will settle a few inches. The amount of compression depends on the quality of the bales. If you are using all-thread, walls can be precompressed by tightening the nuts at your top plate. A torque wrench helps ensure even compression. Precompression makes a wall immediately ready for plastering. Buying tightly compacted bales is advisable in either case.

3-33: External pins and strapping need to be placed in notches and cinched to the wall to facilitate plastering.

Another option is to build the roof, attach it to the top plate, then take a four- to eight-week vacation or work on other parts of the project while the walls settle. The weight of the roof will compress the straw bale walls.

PREPARING WOOD FOR PLASTER. In some designs, plaster is applied over wood. If wood that's to be plastered is only 2 - 4 inches (5 – 10 cm) wide, plaster will go on and stay on well, bridged from one straw bale section to another, so long as you have packed the crevices between the wood and straw bales well (a process we'll describe in Chapter 6). When spanning a surface such as this, be sure you key the plaster into the straw bale wall then bridge the wood.

This system actually provides a more stable and durable plaster base than metal lath attached to wood. The straw in the plaster provides a strong fiber that aids in the bridging effect, tying one section of the wall to the next. In cases with larger spans, you will need to apply a natural adhesion coat to the surface. An adhesion coat creates a gritty surface — a surface with "tooth" — that the plasters key into.

When building a nonload-bearing wall, exposed wood (in the frame, for example) that is going to have plaster applied to it should be coated with an adhesion layer. The adhesion coat consists of a mixture of sand, manure (optional), and flour paste that's brushed onto the wood surface. When it dries, it serves two functions: it coats the wood, preventing moisture from being drawn too quickly out of the plaster (which can cause cracking), and it provides a rough surface to which the plaster will adhere.

Adhesion coats work for many different surfaces, including larger posts, OSB, plywood, metal, cement, tires, and Earthbags. As we'll explain in Chapter 6, plasters applied to wide expanses — 24 inches (60 cm) or more of sheathing — on the exterior may not hold up unless extremely well protected. So when you are designing a house, remember it is better not to apply exterior earthen plasters over sheathing or cement materials that cover a large area. In interior applications, earthen plasters over large spans of sheathing can work well, but only in low humidity areas. In other words, don't apply earthen plasters over sheathing in a bathroom!

In some situations, screws can be used to create an anchor for plaster. Burlap, jute, or reed matting can also be used to produce a rough surface to which plasters will adhere. They are applied to wood and other flat surfaces with an adhesive or flour paste and staples.

In some situations you may need to use metal lath, in which case, you should seal the surface with an adhesion coat first, then apply the lath securely. Lath is used when plastering areas that are not surrounded by straw or earthen walls that can be used to form a plaster span. Because lath does not provide a good long-term stable base, it is best to design a structure eliminating the need to plaster these types of details. Often these areas are finished out in wood, tile, stone, or some other material or are eliminated.

Quick Recap:
Siting and Designing a House for Natural Materials and Earthen Plaster

Designing a house requires common sense and serious consideration to the design details we've outlined above. First of all, remember that your goal is to keep water away from the foundation and the walls. That means you must choose a site wisely. Avoid those

3-34: A layer of plastic sill seal (pink or gray plastic) is sandwiched between the foundation and the sill plate to prevent moisture from wicking into the wood. The bottom plate serves as a moisture break that adds another level of protection. The grade beam is coated with a non-toxic sealer.

CEDAR ROSE GUELBERTH

alluring pieces of property along meandering streams and trout-filled rivers — or at least stay out of the flood plains. In desert country, avoid arroyos — no matter how dry they may seem and no matter how long it has been since they've last carried water, they will some day transport water. If your house is in the way, it will be carried away, too.

Next, study paths of natural drainage on your property, and avoid them, if you can. If you can't, find ways to divert water around the house. Whenever possible, choose a sloped site.

Be sure to build a foundation that keeps the walls of your house dry, no matter what Mother Nature sends your way. Drainage is important. Rubble trench foundations are a great option in many locations. French drains are a good idea, too. Be sure to slope the finished grade away from the house — you don't want water to accumulate next to or underneath your foundation. And be sure to raise the bales and earthen walls off the ground by using a grade beam or stem wall. Design your home for your climate: in deserts of Arizona, your foundation doesn't need to be as deep, nor do your walls need to be raised as high as in a wetter or more severe climate. Further protect the bales or earthen walls from moisture that reaches the foundation by sealing the grade beam or stem wall with a nontoxic sealant and by providing moisture breaks (wooden bottom plate). Don't use moisture barriers such as plastic or roofing felt on stem walls. And don't run plaster to the ground.

Design your roof and overhang so rain won't wash your plaster away, exposing your wall to weather and resulting in deterioration. Trees and bushes can provide a barrier against weather, but don't let them get too close to the building. Gutters and downspouts help protect walls and transport water away from a house but don't count on them exclusively. Gutters can overflow. A rock facing along the foundation and lower walls of a home helps protect it from backsplash. A bed of gravel around the perimeter of the foundation will reduce backsplash, too.

The ultimate success of a straw balestraw bale or earthen wall depends on addressing water issues. Remember: we're not building disposable homes; we don't want structures that will last 50 years, but rather buildings that will still be standing in hundreds of years. Don't be swayed by stories of a friend or neighbor who built a straw bale home five years ago and

CEDAR ROSE GUELBERTH

3-35: Buyer Beware! The design of this house has resulted in a roof that abuts an earth-plastered straw bale wall. The junction of the wall with the roof line is vulnerable to the effects of weather. Even though flashing has been installed along the seam, rain and snow accumulation can easily cause moisture to seep into the abutting straw bale wall and can damage plaster and the underlying wall, resulting in costly repair.

skimped on the foundation or some other important detail; they may be able to get away with short cuts in the short-term, but in the long term, expedient ways often carry a huge price tag. Simply put, a poorly designed, shoddily built home won't hold up as long as a well-designed home built with attention to details. Extra precautions and redundancies are well-worth the additional time and cost. And lest we forget, building a structure that outlasts its cheap imitations also helps to reduce pressure on the world's diminishing resource base.

CHAPTER 4

Understanding Earthen Plasters:

A Closer Look at Their Components

THE SUCCESS OF AN EARTHEN PLASTER on a straw bale or other natural home depends on many factors. As pointed out in Chapter 3, the design of a home is crucial to its success, as is its construction. Properly preparing the walls is essential as well.

The durability of an earthen plaster also depends on the quality of the mix. The quality of the mix, in turn, is determined by the use of proper materials in proper ratios. As we have pointed out earlier, each building site is characterized by unique soil conditions, meaning distinctive types of clay, silt, sand, and aggregate in proportions unique to the location. Because of this, recipes for plaster change from one building site to the next.

How do you know what recipe will work for your project? Fact is you don't. You have to determine your own recipe on a case-by-case basis.

Don't be dismayed, however; simple tests and a little experimentation with potential mixes will allow you to determine a suitable recipe. Before you can proceed to this stage, however, you need to know more about the components of plaster.

> ### FACTORS THAT INFLUENCE A PLASTER'S PERFORMANCE
>
> Design of the house
> Construction and preparation of the wall
> Quality of the plaster mix
> How plasters are mixed
> How plasters are applied
> How plasters are maintained

In this chapter, we'll describe the basic components of an earthen plaster and explain the functions each one performs. We'll examine clay and clay-soils, sand, fiber, and silt, addressing the qualities you'll be looking for in each. We'll also discuss natural additives — substances added to plaster to enhance its workability, application, and performance.

This discussion will provide information that will come in handy when the time comes to make your own plasters. As we've mentioned earlier in the book, our goal is to provide you with a solid understanding of plaster and its components, enabling you to create recipes that work well at your site.

What are the Primary ingredients of an Earthen Plaster?

The three primary components of earthen plasters are clay-dirt, sand, and fiber. Many mudders augment their plasters by mixing in various natural additives such as cooked flour paste or manure. Water, of course, is used to make the mix workable. Water evaporates from the plaster after application, leaving a hard, durable wall finish. Understanding the function of each of the main components, as well as the function of the various additives, will help you create a plaster mix that works well on your home, providing years of unrivaled beauty and protection.

Clay: The Binding Agent of an Earthen Plaster

In Chapter 2, you learned that all plasters must have a binding agent. In earthen plasters, clay serves this purpose.

WHAT IS CLAY? Clay is a material consisting of extremely fine particles. But like many things in our world, clay is a fairly complex substance. You could spend a lifetime trying to understand this diverse and complicated material.

Most clays consist of hydrous aluminum silicates — which means they contain aluminum silicate with associated water molecules. However, there is considerable chemical variation among clays. These chemical differences result in variation in the properties of different clay "species."

CLAY IS A BINDING AGENT AND WATER PROTECTOR IN EARTHEN PLASTERS. Being rather sticky, clay binds to the sand and straw in earthen plaster, and thus holds the mix together. Clay also binds the plaster to walls.

As we noted in Chapter 2, clay molecules often consist of microscopic plates, known as platelets. Water in clay forms a chemical bridge between them, allowing the platelets to bond to one another (see Fig. 2-10). According to soil scientists, the bonding of water between clay platelets retards the movement of free water molecules deeper into the clay. As a result, when clay is exposed to water, it self-seals, creating a water-resistant surface. This helps prevent liquid moisture from penetrating a wall and damaging its interior. Remember, when we say that clay is "water resistant," we're not saying it is waterproof like a sheet of plastic or a rubber rain coat. Rather, clay resists water penetration. When the surface layer of an earthen plaster becomes damp, it retards any further water penetration. Therefore we use the terms *water resistant* and *water protective* when we talk about clay. At the same time, however, clay allows water vapor to pass through. In addition, clay also acts as a preservative, by wicking water away from straw. It also dries quickly, which helps reduce moisture migration into walls, and prevents damage to the interior components of the wall. (The properties are explained in detail in Chapter 2.)

WHERE DOES CLAY COME FROM? Clay consists of fine particles of aluminum silicate combined with natural minerals — some of which color the clays they're in. Clays are found all over the world and come from many different sources. They're often derived from the slow weathering of feldspar, a type of igneous rock which is one of the most common of all minerals within the Earth's crust. Clays are also derived from several other minerals, such as mica.

COHESION OR STICKINESS. Clays exhibit a property soil scientists call *cohesion*. We call it stickiness or adhesiveness — the property which makes clay an excellent binder. Scientists attribute the cohesiveness of clay to the chemical attraction formed between the clay particles and water molecules held between the platelets. (For those with a background in chemistry, it is the hydrogen bonds between clay platelets and water that provide the cohesiveness.)

As you work with various clays, you will find that some are more cohesive than others. Kaolin clays, for instance, are much less cohesive than the smectites, a large group of silicate clays. (Note, however, just because a clay is highly cohesive doesn't mean that it is suitable for a clay plaster. Bentonite, for instance, is cohesive, but is highly expansive and thus not suitable for an earthen plaster.)

PLASTICITY OR PLIABILITY. Clay also exhibits a property soil scientists call plasticity or pliability. Plasticity means that clay is capable of being shaped or molded. This property is most likely due to the plate-like nature of clay particles combined with the lubricating and binding influence of water between the platelets.

Clay is pliable only when wet. However, its plasticity varies over a range of moisture levels. When a little moisture is added, clay becomes slightly plastic. As more is added, it becomes very plastic and quite workable. When too wet, it loses its plasticity. You'll be able to use this knowledge as you work with clay plasters.

CLAY EXPANDS WHEN WET. Clay platelets bind to water molecules. This, in turn, causes clay to expand. When clay dries, it shrinks and tends to crack. This property allows for easy identification in the field. Dry clay also tends to be hard and brittle. However, in a plaster in combination with other components, dried clay retains its superior bond strength — meaning it functions as an excellent binder. Sand and straw in the clay plaster help eliminate cracking.

4-1: Clay like this mud along the Colorado River shrinks when it dries, which leads to cracking. Too much clay in an earthen plaster will cause excessive cracking.

DAN CHIRAS

CLAY IS SLIPPERY. Anyone who's ever handled clay knows that this is a rather slippery material. This distinguishing feature, useful for identifying clay deposits in the field, results from the fact that the flat platelets slide over one another like plates of glass with water between them.

THE COLORFUL WORLD OF CLAY. As Michael Smith writes in his book, *The Cobber's Companion*, "Clay takes on a rainbow of colors." Mineral oxides are responsible for the variety of colors in clay, a gift from the Earth that Mudders have long made use of in finish coats and clay paints to give walls a stunning appearance.

LOCATING CLAY

When driving along highways, you may be able to spot clay deposits by their rich and varied colors. (Although colored soil does not always mean it's clay!) Some clay is brick red. (In fact, that's where red bricks get their color!) Some clay is kind of yellowish, a color often referred to as ochre; others are brown or bluish-gray. If you start looking around, especially in the painted deserts of the Southwest, you'll be treated to a rainbow of clay colors: green, mauve, rose, white, yellow, brown, and many other hues. It is truly one of nature's marvels.

Clay is found along roadways where subsoil has been exposed by bulldozers. It is commonly found around wetlands, although digging here is not recommended for ecological reasons. Check sections of dirt roads that always get slippery after rainstorms, and the places where puddles appear first and disappear last. They appear first and disappear last because clay prevents water from percolating downward. Puddles only disappear when the water evaporates.

4-2: Subsoil usually contains higher levels of clay and lower levels of unwanted organic matter than topsoil. It is, therefore, usually the best source of material suitable for earthen plasters.

YOUR MAIN SOURCE OF CLAY: THE SUBSOIL. Clay is widely dispersed on Earth. It is found in both the topsoil and subsoil. Clays are also found in deep deposits in the Earth's crust, on the floors of lakes and oceans, in certain glacial deposits, in desert basins, river deltas, in windblown deposits (called *loess*), and in a number of less significant areas.

Although clay is present in the topsoil in most areas, greater concentrations usually occur in the subsoil. Moreover, topsoil contains organic matter that has no place in an earthen plaster. Organic matter consists of live plants, roots, and bits and pieces of plants and dead animals, such as earthworms and insects, in various stages of decomposition. Organic matter may decompose in a plaster and has little binding capacity. As a result, it can compromise the integrity of an earthen plaster. Organic debris in a plaster also makes application more difficult. Subsoil generally has a much lower content of organic material. The occasional roots that may occur in the subsoil can be screened out or removed by hand.

Because of these factors, most mudders use subsoil to make their plaster, the layer of dirt beneath the topsoil. The only exception is in desert climates. Here it is often possible to collect clay on the surface as there is little, if any, topsoil.

Because you may be using topsoil or subsoil to make plaster, we'll frequently refer to your starting material as clay-rich dirt, clay-rich soil, or just plain dirt. Note, too, that clay-rich dirt contains other things as well that are important to a plaster — for example, some silt and sand (both described shortly).

HOW MUCH CLAY IS IN YOUR SUBSOIL? Determining the amount of clay in soil requires a few simple tests. Be sure to run these tests long before the plastering begins, preferably while you're in the design phase of a project. Don't wait till the last minute. If your soil flunks the tests, you'll need to augment it or import a clay-rich dirt from another site. If your plaster crew is ready to go and your plaster mix needs enhancement, you'll end up wasting their time while you are trying to "fix" the dirt. And time, of course, can create a monetary penalty on building sites. If your dirt "passes" all the tests described below, it probably has a decent clay content.

So how do we go about testing a site for the presence of clay?

GRAB A HANDFUL. First dig down through the topsoil on your site; it's usually dark and fairly soft. Beneath the topsoil is the subsoil, a lighter-colored material that is usually much harder and more difficult to dig through. Once you reach the subsoil, scoop some loosened dirt into your hand. If it is moist, compress the material in your hand, then squeeze it tightly. If the subsoil is dry, slightly moisten it, then squeeze. If the dirt contains clay, it should bind together and hold its form. The better it "holds together," the higher the clay content .

THE WORM TEST. Next, perform a simple worm test — so named because you'll be making a pencil-thick worm out of earth. We'll call it a "dirt worm" so you won't get confused with the real thing, the annelid earthworm.

First put a little of your dirt in your hands. If it is dry, spit into the dirt or moisten it by sprinkling water on it, then mix it in your fingers. When the sample is thoroughly mixed, roll it back and forth to create a pencil-thick dirt worm. As you roll the moistened soil around, you'll begin to get a feel for the mix. Pick it up. Does your dirt worm remain intact? If so, your clay content is probably sufficient. If it falls apart, you may need to add more water and if that doesn't work you may need to supplement your subsoil — that is, add some more clay.

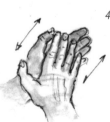

4-3: The Worm Test

GET THE FEEL FOR IT. Next, take a larger amount of dirt into your hand and wet it down. Get it moist. Play with it a bit. Pull it apart. Roll it into a ball and make a patty cake. How sticky does it feel? If the dirt isn't sticky, you may need more water; wet it slightly and work it again. If it still isn't cohesive, don't despair; you can add some powdered clay, flour paste, manure, or other natural additives to increase its cohesiveness.

JAR TEST. This allows you to determine the proportions of clay, silt, and sand in your dirt.

First, fill a clear glass jar one-third full with dirt. A one-quart or one-pint mason jar (or any empty pickle jar) will work. Pick out small rocks and organic matter, such as roots. Break up any clumps in the subsoil. If they are difficult to crush, try adding a little water and letting them soak for a few hours.

Next, add water, filling the jar two-thirds of the way to the top. If you add a teaspoon of salt to the mix, it will speed up the deposition of clay, which as you'll soon see is the slow poke in the group. Next, place the lid on the jar and screw it on tightly. Shake the jar vigorously for a minute or two over a sink or outside in case it leaks. Alternatively, if you have an old blender, you can dump the mix into the machine and let it do the work. Let it run until the test sample is thoroughly mixed. When thoroughly mixed, pour the contents of the blender into a jar.

After shaking or blending, place the jar on a flat, level surface. The components of the mix will begin to settle out. The largest, heaviest particles in the dirt (pebbles and coarse sand) fall to the bottom within seconds. After about three seconds, mark the side of the jar to indicate the top of the sand and pebble layer. Finer sand particles settle out next, then silt: both should settle out within ten to fifteen minutes. However, the line between silt and fine sand is sometimes difficult to distinguish. Michael Smith, a veteran cob builder, draws the line between silt and sand when he can no longer see individual particles.

4-4 (a) (b): The jar test. (a) Soil to be tested is placed in a jar. Water is added and the jar is covered and shaken. (b) After vigorous shaking, the jar is placed on a flat surface and the materials begin to settle out. Measuring the layers tells you how much sand, silt, and clay are in your soil.

Clay particles, which are even finer than silt, settle out last. They're so fine, in fact, that they may remain suspended in solution for several days. If the water in the jar turns clear quickly, that's a sign that there's not much clay in your dirt. If it remains cloudy or opaque after the sand and silt have settled out, you're in luck: there's clay in your subsoil. You may also note that some organic matter may be floating on the surface of the water;

it won't cause any problems. After the clay settles, measure the thickness of each layer to determine the relative proportions of clay, silt, and sand.

The jar test may take a few days, or even a week, to complete — and that's another reason why you should test your soil well in advance of plastering. The jar test provides a reasonably good estimate of how much clay, silt, and sand are in a dirt sample. However, because clay expands when wet, you can achieve a more accurate measurement by drying the sample. After the clay settles out, scoop out the clean water on top of it, then let the jar sit with the top off. Put it out in the sun to accelerate the drying. When dried, you can make a more accurate measurement of the clay content and thus more precisely determine the ratio of clay to silt and sand.

An earthen plaster soil should contain 5% to 12% pure clay, according to Gernot Minke, author of *Earth Construction Handbook* and a professor of architecture as well as the director of the Building Research Institute of the University of Kassel. The ratios of clay to sand to silt — and the characteristics of the clay — will determine what you need to do, if anything, to make your soil suitable for plaster.

TESTING THE SOIL AT SEVERAL LOCATIONS. When performing tests, you may want (or need) to sample subsoil from several locations on your building site. It is usually more convenient and more economical to use dirt that's been excavated for a foundation or a basement, rather than excavating and transporting suitable dirt from other locations. However, you may find a better soil thirty feet away from the foundation excavation. Subsoils can vary quite considerably within a relatively small distances. Finding a clay-rich subsoil may require you to dig a couple of test holes first. By testing several locations at a building site, you increase your options. It's great to have several choices.

WHAT DO YOU DO IF YOUR SOIL DOESN'T CONTAIN ENOUGH CLAY? If your dirt is clay-poor, and therefore lacks cohesion and other properties needed for a successful plaster, you can amend it. One option is to mix in a higher-clay-content dirt from another source, preferably one nearby. If you don't have a source, you can always augment the subsoil on your site by purchasing high-clay

BEYOND RATIOS

Ratios provide you with important information — notably, how much clay, sand, and silt are in your starting material. This will help you decide which components need to be added to your mix to enhance its performance. However, ratios do not determine the quality of various components of a plaster mix, most notably the clay. Clays vary chemically and chemical differences result in differences in properties: pliability, cohesiveness, and so on. Playing around with your soil, as described earlier, and experimenting with it are both critical in determining the performance of soil in plasters.

TAKE TIME TO FEEL THE PLASTER

If you attend a workshop on plastering, be sure to take some time to feel the various components. Run your hands through the soil that is going to be used to make plaster. Moisten it, roll it around in your hands, and assess its stickiness. Do the same for the various plaster mixes. Close your eyes and really concentrate on the feel of things. That way, when you get home you will have a good sense of what your plaster should be like.

INCREASING THE ADHESIVENESS OF A PLASTER MIX: OPTIONS

1. Blend subsoil with a higher clay content subsoil
2. Extract clay from subsoil and add it to your mix
3. Add supplements such as manure and flour paste
4. Add bagged clay

4-5: Workers apply an earthen plaster to St. Francis Church. Their soil contained a suitable ratio of clay, silt, and sand and could be used as is with water and straw added.

CEDAR ROSE GUELBERTH

content dirt from a local supplier. Overburden (the dirt on top of a gravel deposit) may be suitable: it can be obtained from local gravel pits and is usually quite inexpensive. Excess dirt from nearby construction sites may also be suitable. Be sure to test any soil you are considering for use in plasters. However, you'll quickly discover that while dirt is cheap, having it hauled to your site is not. But even so, for a few hundred dollars, you can often obtain enough clay-rich subsoil for a complete project.

You can also amend soil by adding bagged (dry, powdered) clay, such as kaolin, which is available at local pottery supply stores or other outlets. Flour paste and manure added to plaster also increases the stickiness and pliability of a subsoil — essentially reducing the amount of clay you need in your mix. We'll talk more about these options a little later in this chapter and in Chapter 6.

A more labor-intensive but more environmentally benign way of securing additional pure clay is to extract it directly from some of the subsoil on your site. Extracted clay is added to your subsoil to increase its content. Harvesting clay from dirt is relatively simple: To begin, screen clay dirt over a tarp, wheel barrow, large tub, or garbage can to remove rocks. A 1/2-inch hardware cloth (screen) works fine. Break up chunks of clay with your hands by rubbing them against the screen.

Fill a tub or trash can one-third full of water, then add screened soil while stirring with a hoe or mixing with a paddle mixer. Continue to mix while adding dirt to prevent clumping.

After the dirt has been thoroughly mixed and clumps of clay have been broken up, give it a few minutes to allow sand and aggregate to settle. You can now begin to pour off or scoop out the material that forms on the top. Called *clay slip*, it consists of clay and some silt and can be added to your plaster mix to increase its clay content.

If you have more time, let the mixture settle for a day or two, remove the water on top, then carefully scoop out the layer of clay that forms on the top. Alternatively, you can wait until the mixture dries out, then remove the clay from the top.

If you need to substantially increase the clay content of your soil, locate a local source of a higher clay-content soil, and use it either to mix into your existing soil or to use in place of existing soil. Alternatively, you can add some dry, powdered (bagged) clay to increase the clay content of your soil. You may also be able to increase the adhesiveness of your existing soil by using natural additives (such as flour paste and manure).

CEDAR ROSE GUELBERTH

4-6: To extract clay from subsoil, fill a tub or durable garbage can one-third with water, add screened soil, and mix thoroughly with a hoe or paddle mixer. Clay will be suspended in the water on top. Sand and aggregate will settle to the bottom quickly.

Sand: Structural Strength

Sand consists of tiny mineral granules of rock, its parent material. Predominantly composed of quartz, or silicon dioxide, sand is a chemically non-reactive substance.

SIZE, SHAPE, AND HARDNESS. If you've spent much time on beaches, along rivers, lakes, dunes, and the world's oceans, you know that sand varies in color and grain size: some sands are very fine, others are quite coarse. In fact, sand grains vary from 0.05 mm to 2 mm in diameter.

Grains of sand also vary in shape: Some are quite angular and sharp; others, like beach sand, are often smooth and round, having been polished by waves for centuries. Dune sand tends to be fine-grained and smooth, after having been polished when the wind blows sand grains over one another. Sand also varies in hardness: some sand grains like quartz sands are very hard; others are less so.

WHY ADD SAND TO A PLASTER? In an earthen plaster, sand provides structure, strength, and bulk. In a plaster, sand particles are bound together by clay. On its own, clay does not create the structure required for a good plaster. It's not terribly stable, but rather pliable. Sand, on the other hand, is geometrically stable. It transfers this property to an earthen plaster. Combined, clay and sand make a strong and stable material.

4-7: Carole Crews and her kids sift beach sand for use in plasters. If you collect sand from naturally occurring deposits, especially beaches, we suggest you rinse it before use to prevent cloudy salt deposits which can migrate to the surface of your finish.

Sand provides other benefits as well. For example, it doesn't expand or contract when it is wet, as clay does. This helps stabilize an earthen plaster, preventing it from cracking. Sand is also water insoluble: it won't dissolve away. It is also chemically nonreactive: it won't react with clay or other components of an earthen plaster.

USING WHAT'S AVAILABLE OR ADDING SAND. Because sand occurs naturally in many subsoils, it may not be necessary to add any sand at all to your dirt to make plaster. If there's not enough sand in your subsoil, however, you will need to add some.

Sand can be obtained from a number of different sources. You can, for example, harvest it along nearby river banks or at local beaches. (Beach sand containing salt may cause white blotches to form on the wall, as the salts migrate out of the plaster, so be sure to wash it first.) If natural sources aren't available, you can purchase sand by the truckload from local sand or gravel pits. Either way, local sand should generally be run through an 1/8-inch metal screen to remove small stones and organic debris.

If you elect to purchase sand from a local source, such as a gravel pit or sand pit, we recommend buying washed masonry sand, which can be purchased in dump truck quantities. Masonry sand is also relatively inexpensive. Washed masonry sand is clean (free of dirt or silt) and works well in earthen plasters and generally will only need to be screened if it is used for finish plaster (final coat).

Washed masonry sand is often suitable for finish plasters. If you want a finer finish coat, silica sand can be used, though this is not necessary. As a general rule, coarser sands are used for the first layers of plaster, and finer sand for finish coats.

Of the various sources of sand, the most expensive is bagged sand, which is available at many building supply stores. Be careful not to purchase play sand or any similar product, as they often contain a lot of dirt and silt.

Fiber: Providing Tensile Strength and Reinforcement

Fiber is the third ingredient of plaster. Dry straw, hemp fiber, cattails, and animal hair (such as wool) are all suitable. Horse manure and coconut and sisal fibers can be used as well. Straw is the most commonly used fiber because it is widely available, easy to work with, effective, and relatively inexpensive.

WHY ADD FIBER? Fiber forms a reinforcing meshwork in plasters, which helps to hold the mix together when it is wet as well as dry. Fiber also provides some flexibility in a dried plaster. As you may recall, when clay dries it shrinks and tends to crack. This cracking can be countered by fiber, as well as sand. If your mix is good, you should have little, if any, cracking.

DRY, MOLD-FREE FIBER: A MUST. The fiber used in plasters must be clean (not moldy) and dry. When ordering straw, a commonly used material for plasters, be sure it hasn't gotten wet. Open up a few bales and look for discoloration — sure evidence of mold. Mold usually appears dark, but is sometimes white or other colors (yellow or gray). A clean, dry straw should have a golden color throughout. You can also smell the straw to see if you can detect the odors given off by mold. You can test bales with a moisture meter prior to purchasing or use: high moisture levels may indicate the presence of mold.

When the straw is delivered to your site, store it in a dry place off the ground. If you come across any wet or discolored straw, dump it in your compost pile or use it to mulch your garden.

WHAT SIZE FIBER? Add fiber of different lengths to earthen plasters. Variations in length of fiber creates a stronger material.

In the first two layers of plaster (the base coat), larger fibers, measuring 4 - 8 inches (5 - 10 cm), are appropriate, but you should also throw in some 1- to 2-inch (2.5 - 5 cm) fibers. The final or finish coat does not require large-diameter or long fibers. They can compromise the integrity of an exterior finish coat and may lead to surface deterioration.

> ## VARIETY IS THE SPICE OF LIFE
>
> Add fiber of different lengths to earthen plasters. Variations in length and type of fiber (straw, cattails, wool, etc.) create a stronger material.

If you like the appearance and want to add straw to a finish coat, be sure to chop it to lengths of 1 - 2 inches (2.5 - 5 cm). While eliminating straw in the final exterior coat is recommended, you may want to add other forms of fiber. Adding manure, for example, provides microfibers that improve the quality of a finish coat, as do the small fibers of cattail fluff and wool. (For more on fibers in plasters, see Chapter 6.)

ACQUIRING STRAW FOR PLASTER. Because straw is the most widely used fiber for earthen plasters, we'll focus our discussion on this product. Bear in mind, though, that other fibers work well, too.

Straw for plaster can be acquired from several sources. When building straw bale homes, you will find that straw of varying lengths accumulates around your pile of bales and along the walls of your home. Loose straw will be especially abundant after walls have been shaped and trimmed. Splitting bales produces loose straw, too. We recommend that you gather up loose straw as construction proceeds. Make sure it is clean (that is, free of debris) and kept dry. Store it under tarps or in burlap or other bags. If possible, store indoors. Do not store it in closed plastic bags; moisture can accumulate in such instances. Gathering straw before it gets wet or dirty not only provides a source of fiber for plasters, it helps to reduce potential fire danger.

CHOPPING STRAW. For a successful plaster, most straw needs to be chopped into smaller lengths. Occasionally, you'll find straw bales containing short strands. Many times this is an indicator of poor-quality straw bales. Although they're not suitable for building walls, they can be great for making plasters as long as the straw is clean and dry.

There are a number of ways to chop straw. Many people throw straw into a clean garbage can, then have at it with a weed eater or weed whacker. If you have access to a leaf and grass mulcher, straw can be chopped this way as well. Straw can also be shredded with a chain saw.

Some mudders grate straw on a ¼-inch wire screen or metal lath. To do this, grab a handful of straw, then rub it along the surface of the wire. Be sure to wear gloves to protect your hands. Smaller pieces will fall through the grate. You can even grate straw successfully over diamond lath draped over a cardboard box. (Neither process is very fast and therefore not practical for larger projects.)

OTHER SOURCES OF FIBER. Natural fibers can be collected from a host of other sources. Cattails, for instance, are a great source of fiber for plasters, especially finish plasters. They're collected in the fall when the cattail heads are brown and ready to let loose their small fluffy fibers.

Wool and hemp fibers are also good for plasters. Raw wool can be teased apart and cut into more manageable lengths with a pair of sharp scissors. Manure from a variety of animals works well, too. Fresh horse and cow manure can be collected from nearby farms and transported to your site. Horse manure is a better source of fibers whereas cow manure is a better source of enzymes that improve the adhesive qualities of a plaster. However, either one will do.

Additives: Improving the Durability and Function of Plasters

Most earthen plasters consist of clay-rich dirt, sand, and fiber mixed with water. As noted earlier, additives are often used to improve the quality of a plaster: they add strength and produce a more cohesive, durable plaster. Whether you add them to a finish coat or all coats depends on the quality of your raw materials. If you're starting with sandy or silty dirt with low clay content which isn't very sticky, you may need to supplement all mixes, from the base coat to the finish coat.

Some of the most common additives are cooked flour paste, manure, cactus juice, casein (milk protein), and various natural oils such as linseed oil and canola oil. Some other additives include alum, natural glues, gum arabic, kelp, lime, and powdered milk. Salt, stearate, tallow, tannin, leaves and bark of certain trees, and xantham gum have also been used. Ox blood has been used in some cultures and is still used in traditional cultures for adobe floors as well as plasters; the protein in the blood is believed to create a more durable finish.

From the list of additives shown in sidebar, you can see that there are many ways to achieve the plaster qualities you need. There are also many additives we haven't included, which are unique to local cultures and determined by local availability. We'll focus on the main additives in this book to keep things simple and help build your understanding of plasters.

FLOUR PASTE. Cooked flour paste is an easy-to-make, readily available material made from water and flour. In a plaster, flour paste serves as a binding agent and a hardener. That is, it makes plaster more adhesive when wet. It also results in a harder, more durable, and dust-free finish when dry. Flour paste also makes plaster smoother and improves its "workability" — that is, how well it goes on straw bale and earthen walls. (We'll describe how flour paste is made in Chapter 6.)

Some people are concerned about insects and rodents eating their walls or mold forming on them. However, the amount of flour paste added to plaster is so minimal it does not attract critters or mold. Cooked flour paste has been used successfully for centuries in a number of products such as wall paper glue and plasters — without molding.

MANURE. Manure is also an effective additive: it serves as binding agent and gives plaster more body. The binding properties of manure most likely come from enzymes and proteins in the material, which make plaster stickier when wet and more water-resistant when dry. Manure also contains small fibers that strengthen a plaster and reduce cracking and water erosion.

Manure probably first began to be added to plaster by accident. In earlier days, draft animals were often used to mix plaster. Tethered to a center pole, horses were driven around and around over a plaster mix: their hooves mixed the plaster and their intestinal tracts supplied free additives. Early mudders liked the results.

NATURAL ADDITIVES

The following additives increase the workability, durability, and water resistance of an earthen plaster, and can minimize dusting:

- Cactus juice
- Cooked flour paste
- Casein or milk powder
- Oils

- Alum
- Kelp
- Lime
- Gum arabic

These additives increase the pliability and strength of plasters by adding fiber:

- Animal hair
- Cellulose (newsprint)
- Cattails

- Cow manure
- Horse manure
- Leaves and bark of certain trees

4-8: Historically, horses tethered to a center pole were used to mix plasters. Manure dropped into the plaster improved the quality of the finish product. Modern plasterers often add manure for the same reason.

Many different types of manure have been tried over the course of time, each with its own unique benefits. Horse manure, for instance, has a high microfiber content. Cow manure has more microbes and enzymes than horse manure but has a lower visible fiber content. (The reason cow manure contains fewer fibers is that cows are able to digest cellulose, thanks to special bacteria that live in one of their four stomachs. These bacteria contain the enzymes needed to digest cellulose in the grasses they eat, so goodbye fiber!) Some people have reported success with llama and alpaca dung.

Manure should be as fresh as possible: composted or decomposed manure loses its enzymes and its adhesive quality and fibers break down. Manure should be sifted before use and can be used instead of chopped straw in finish plasters. You'll need to experiment to determine how much to add.

CACTUS PRICKLY PEAR. Prickly pear cactus "juice" is one of the most commonly used additives. Part of the reason for this is that it is one of the most prevalent species of cactus in the world: it can be found in deserts, grasslands, and some open forested areas. It grows in 48 of the 50 states in the United States, and in other parts of the world, such as Australia. In fact, in Australia prickly pear cactus became a nuisance, as it overran many parts of the country. Prickly pear cactus is also quite prolific in other parts of the world. In such instances, harvesting some won't drive the species into extinction. Some species produce large leaves, too, which yield lots of material. One of the best prickly pear species, known as *Opuntia ficus indica*, grows tall throughout the deserts of the U.S. Southwest and Mexico. It produces few large spines, which make it easy to work with, and contains lots of pectin (defined below).

The juice from the prickly pear cactus leaf pads and the stems of the cholla cactus serve many functions. According to some sources, they help a plaster set, increase its stickiness or adhesion, and they improve its workability. Cactus juice, as we call it, also serves as a stabilizer — that is, it helps make earthen plasters more water resistant and more durable. It also prevents dusting.

Cactus juice works well because it contains a water-soluble long-chain carbohydrate — a complex sugar or polysaccharide. According to veteran straw baler and mudder Bill Steen, this gooey substance, known as *pectin*, is a binding agent that increases the adhesion of an earthen plaster. Pectin is also responsible for increasing the water resistance of earthen plaster and has been used to augment lime plasters in both Mexico and the southwestern United States for hundreds of years.

4-9: The juice from the leaf pads and stems of cacti increases the stickiness or adhesion of a plaster, improves its workability, and serves as a stabilizer, making the plaster more water resistant and more durable.

CEDAR ROSE GUELBERTH

In some locations, prickly pear cacti are just too small to work with. The amount of pectin varies considerably among the different species of prickly pear cactus, too. In such cases, you will need lots of material to create sufficient additive. You're better off using cooked flour paste, which performs a similar function and is easy to prepare.

Oils. Various oils can be added to earthen plaster to enhance its water resistance and durability. Linseed oil is one of the best but is used only for finish plaster coats. When exposed to oxygen in air as the plaster dries, linseed oil becomes oxidized, and is then "fixed" in the wall. It also increases the flexibility of a finish plaster, allowing it to expand and contract more than untreated plaster, which helps to reduce cracking. Linseed oil does all this while continuing to allow wall to breathe — that is, it permits water vapor to pass through.

Be careful, however: too much oil in your plaster can break down exterior finishes. How much should you use? The following are useful guidelines: use about 0.5 to 1.0 percent (by volume) linseed oil in the finish coat, or try a tablespoon of linseed oil mixed in a five-gallon bucket of plaster.

> ## EXTRACTING CACTUS JUICE
>
> Approximately two weeks before plastering is to begin, place chopped prickly pear leaves or cholla stems in a plastic bucket or other airtight container. Fill the container with water. In about one to two weeks, mucilage will be formed. Separate it from the remains of the leaves and use immediately. If mucilage is required more quickly, chop prickly pear leaves and cholla stems, then immerse in water and boil for a while. Mucilage can then be separated out. You'll need to experiment to determine the amount to use in your plaster mix. We recommend starting with one cup in five gallons of plaster.

A variety of other oils can be added as well, including canola oil, hemp oil, and flax seed oil. However, these have less body than linseed oil — that is, they are less viscous (thick). The more viscous linseed oil also adds pliability, which other oils don't.

Oils come in variety of commercial preparations. The commercially available products you buy at lumberyards and hardware stores, however, can contain some pretty toxic components, even when the label reads "100% oil." From the perspective of personal and environmental health, the all-natural oil products are far better. Use food grade oils or products from Bioshield, AFM, Auro, Livos, and other suppliers listed in the Resource Guide.

Although oils have been used by many plasterers, we recommend caution. This seemingly good idea can cause disastrous results. In exterior applications, there is a fine line between too much and not enough oil. Small amounts added to a mix slightly increase the durability of a finish coat. Too much oil may appear to provide a well-sealed wall surface in the short term. Over time, however, the oil-treated finish coat will begin to heat and cool as a separate layer, causing the finish coat to separate from the underlying plaster layer, most often in exterior applications.

Another concern is that exposure to weather can cause small openings or pores to form in the surface, allowing moisture to enter the plastered surface of a wall. Moisture

> ## WARNING!
>
> Although oils can be added to produce a beautiful and durable exterior wall finish, we recommend caution. Some people have used oil in finishes to waterproof walls while disregarding the design elements essential to the success of a plaster. Too much oil, and exposure to the sun and weather can actually cause the finish plaster on an exterior wall to break down and peel off. Oil in base layers can prevent good adhesion of one coat of plaster to another. Too much oil in a finish coat can make it difficult to repair finish plaster, too. If you use an oil in your finish coat, remember that more oil does not mean more protection. Less is better!

may become trapped in the plaster layers beneath the finish coat, eventually eating away at the base layers. As a result, the finish layer will peel off the wall. However, oils and waxes can be applied to interior finish coats for a beautiful finish.

Silt vs. Clay: Understanding and Detecting the Crucial Differences

Before we move on, we need to say a word or two about silt. Let's take a few moments to answer three basic questions: What is silt? How can you tell it apart from clay? How does it affect plaster?

WHAT IS SILT? Silt, like sand, is made of tiny particles of rock, which are intermediate in size between sand and clay — that is, they're larger than clay but smaller than sand. Unlike sand, silt particles are much too fine (small) to see with a naked eye.

Silt particles are irregular and diverse in shape, seldom smooth or flat. Some geologists think of silt as "microsand particles" with quartz being the predominant chemical component.

When wet, silt often feels slippery, as clay does. If you moisten some silt and rub it between your fingers, you'll find that it feels a lot like clay. The reason for this, say geologists, is that silt particles often contain a thin film of clay on their surfaces. This gives silt some plasticity and cohesion (stickiness). It is also water-absorbent. All three are properties of clay.

This, however, is where the similarities abruptly end. Clay particles, as we've noted several times earlier, are flat plates. Clay is much more plastic, too, meaning it is highly workable. That's why potters use it and why kids play with it. The worm test we described earlier relies on clay's plasticity. If the clay content of your soil is high enough, you can bend the clay worm around your finger and it won't break — so long as it is wet enough. The clay's plasticity allows the dirt worm to bend easily: the more clay, the more bend. The less clay, the more the worm has a tendency to break at the bend. A soil with a high silt content has much less plasticity: a dirt worm will break apart very easily.

While both clay and silt feel sticky, clay is much stickier. Clay molecules are held to one another by water molecules. This bridging effect gives clay its plasticity as well as its stickiness.

Although silt feels slippery when held between the fingers (because silt particles are coated by a thin film of clay), silt doesn't have much binding power — that is, it is not as adhesive as clay. Too much silt in a plaster may interfere with clay's ability to bind to sand and straw and to other clay molecules.

Although excess silt can physically block clay from binding to sand and straw and therefore may produce a weaker plaster, don't panic if you have silt in your subsoil (which is highly likely!). Some silt is beneficial: clay-dirt containing some silt can produce a highly effective and lovely plaster that is easy to apply and durable. Some of our best plasters have had a fairly high silt content! Note, too, that silty subsoils are easy to amend with natural additives.

TESTING FOR SILT. To assess a soil's silt content, take a pinch of wet dirt between your thumb and index finger and squeeze them together. Then try to pull them apart. If your finger and thumb stick together, there's probably sufficient clay in the soil. If they don't, there's probably less clay and more silt. You may need to add clay or flour paste (or some other additive) or seek another source.

Another test to determine the amount of silt vs. clay is to take a small clump of wet soil (about the size of a golf ball) in your hands. Flatten it against your palm, then turn your hand palm down. If the clump sticks to your palm, you've got more clay and less silt. If it doesn't, your soil probably has more silt.

DISTINGUISHING CLAY FROM SILT IN THE FIELD. Being able to distinguish clay-rich soil from silt-rich soil will come in handy when gathering clays for plasters, finish coats, and aliases (clay paints). Telling clay and silt apart in the field can be quite challenging — which is why it's important to test an imported soil before it is delivered to your site!

You can distinguish between clay- and silt-rich soils by the "stickiness test" we've just mentioned. You can also tell by the way the two harden: when high-clay soil dries, it tends to clump. Clay clumps can be quite hard — almost rock hard! If you try to crumble it in your hands, you're in for some work. Dirt containing silt, on the other hand, crumbles much more easily.

CHARACTERISTICS OF SILT

- Extremely fine particles, invisible to the naked eye
- Slippery when wet
- Water repellent, if in a thick enough layer
- Not as sticky as clay
- Poor adhesive and binding agent
- Impairs water vapor movement when concentrated in a plaster
- Tends to crumble rather than clump when it dries
- Feels like a very fine powder but is gritty when rubbed on your teeth
- Easier to dig through than clay

Clay can make digging difficult. If a soil contains a lot of clay, it is harder to dig in. The soil also tends to come out in clumps. Silty soils, even with high clay content, are easier to dig through and don't clump nearly as much as clay-rich subsoils. Pure clay curls when it dries; silt doesn't. (When silt is evenly distributed throughout a soil, it prevents clay from clumping and makes dirt easier to dig and sift, and sometimes makes an ideal starting material for plaster.)

There are two other physical characteristics that help to distinguish between clay and silt. In loose soil, silt feels like fine flour when dry. Rubbed on your front teeth, it feels gritty. When ground, clay is even more powdery. When you rub clay powder on your front teeth, you won't feel any grittiness.

Now that you know how to distinguish clay from silt, we want to be sure to reiterate an important point: silt does tend to decrease the binding capacity of clay in a plaster, but some silt is okay, and can actually be of benefit. As noted earlier, some of our best earthen plasters have had a fairly high silt content. Clay-dirt with some silt can produce a beautiful material that is workable and easy to apply. So don't panic if you encounter silt in your subsoil.

Are there any Types of Bagged Clays or Clay-Rich Soils You Should Avoid?

Clays come in many varieties, each with its own name. For example, you'll encounter kaolin, a whitish clay, and bentonite, a highly expandable clay. As we noted in Chapter 2, the differences in clay are due to differences in their chemical and physical properties. In other words, each one has a unique chemistry as well as a unique microscopic structure. Each clay behaves differently, too: although all clays are sticky, plastic (workable), and expansive, they differ in these properties. Some are stickier and more expansive than others. Even within the same family of clay — for example, the kaolins — you'll quickly discover variations in clays from one region to another. Georgia kaolin differs from Iowa kaolin, which differs from kaolin from China. Should you avoid any particular types of clay?

Yes. Bentonite and clays with similar properties. Bentonite is one of the worst clays for earthen plasters or clay-based finishes because it is highly expandable. It shouldn't be used because it just expands too much (19 times) and will cause excessive cracking no matter how much sand you add! Bentonite is also so intensely hydrophilic (water-loving) that it quickly absorbs water. Put a drop of water on it, and the bentonite binds the clay so rapidly that the plaster is virtually unworkable. We don't recommend the use of bentonite or other similar clays in earthen plasters or clay-based finishes.

AVAILABLE VS. UNAVAILABLE CLAY. Clay is a remarkable substance. However, some deposits (both underground and surface ones) contain so much hydrophilic clay that they' re extremely difficult to work with. The clay can be so bound up that it is difficult to extract. Once you excavate the stuff, the chunks are rock hard, and nearly unbreakable unless you've got a sledge hammer! In this state, the clay is obviously not suitable for mixing.

If you encounter clay such as this, you will need to harvest it, then soak it in water for a while — as long as it takes to soften it. You'll then have to stomp on it or smash it, then mix it with your hands or some other implement. This will start to break up the clumps; additional soaking will probably be required, and you may have to repeat the process several times until the clay is hydrated and workable. You may want to locate another source of good-quality clay-rich dirt (containing more silt and sand) from a local source, for instance, a gravel pit, in the subsoil, or along on the banks of nearby river or pond where the soaking has already been done for you.

HEALTH WARNING: WEAR A MASK!

Soil, clay, sand, and fiber are all natural materials and relatively safe to use. However, you should be aware of some potential health hazards. Most clays, for example, consist of extremely fine particles of aluminum silicate; kaolin is a good example. Used as an extender in medicines and as an absorbent in diarrhea medications, some consider it to be safe to ingest, although breathing aluminum oxide is clearly not recommended.

Powdered clays are especially troublesome. Clay particles are so fine that whenever a powdered clay is scooped out of a bag, poured from one container to another, or sifted, it produces a dust cloud. Dust is produced by many other activities involved in plastering a home: screening or sifting subsoil, pouring bagged clay, shoveling sand, or chopping up straw for plaster, and adding pigment to mixes. Be sure to take precautions. Dust can be breathed into your lungs and may cause serious illness. To protect yourself against any possible health problems our advice is: always wear a certified dust mask when working with any of these materials in a dry form.

Quick Recap

Well, that's it. You understand that earthen plasters are made of clay-soil, sand, fiber, water, and various additives, and know what each component contributes to the plaster. Clay is the binding agent that holds this complex mix together both when wet and when dry. Sand provides structure and strength. Fiber provides strength and some flexibility. Like sand, fiber prevents cracking. Water allows all of the components to be mixed and applied easily to the wall. Additives provide additional adhesion, which is to say, they assist clay as a binding agent. They can also increase the durability of a plaster and its water resistance.

To make a durable, long-lasting plaster requires the proper ratio of each. As noted earlier, the proportions of these ingredients in your plasters will vary depending on the ingredients themselves. Clays found in subsoils are chemically variable and behave differently. The amount of sand and silt also varies from one site to another.

Because subsoils vary, rely on the feel of a plaster, the way it goes on the wall, and the way it dries to determine suitable recipes (a process we describe in detail in Chapter 6). Test your mixes well. Start with a couple of small projects — for example, a shed or outbuilding — before you plaster a whole house. The experience will pay huge dividends.

Now that you understand the components of plaster, we'll turn our attention to site preparation and mixing earthen plasters.

CHAPTER 5

Preparing Your Work Site and Mixing Plasters

EARTHEN PLASTERS consist of clay-rich dirt, sand, straw, and natural additives. Each component performs a specific function, described in Chapter 4. Now that you've got a good idea of what goes into an earthen plaster and how each component functions, it is time to move to the next stage: preparing your site and mixing plaster. We'll begin our discussion by looking at ways to make a work site safe and highly efficient. We'll then discuss the simple tools that are required to mix plaster. After that, we'll tackle plaster mixing itself. Finally, to help lay the groundwork for Chapter 6, we'll describe each layer of plaster you'll be applying in straw bale and earthen-walled buildings.

Preparing Your Work Site for Efficiency and Safety

Mixing plaster requires a fair amount of space. On a house project, a mixing area must accommodate piles of sand, clay-dirt, straw, wheel barrows, and a mixer.

Organize Your Site for Efficiency

To facilitate a smooth and safe means of mixing plaster and delivering it to walls, a work site should be well-organized. We recommend careful staging of tools and materials for mixing — that is, organizing both tools and materials in a logical manner. Staging will not only make the process efficient, it can substantially reduce the amount of work that's required to mix plaster and deliver it to the walls. Be efficient at this stage. You'll get plenty of exercise applying the plaster!

Staging begins early in the building process, during the planning. Take some time to map out the best site for mixing plasters. As you devise a plan, consider all areas, access by trucks, people (including safety), and proximity to the area being plastered.

For an efficient operation, be sure that the mixing site is as close to the building and interior entrances as possible. This will minimize transportation distance and energy spent moving mud to the walls you will be plastering. Be certain that your piles are

> "Plastering is hard, heavy work. You want to minimize the amount of effort you expend on anything other than applying plaster to the wall, so prepare your site thoroughly."
> — Chris Magwood and Peter Mack
> *Straw Bale Building*

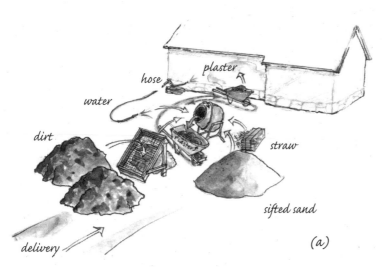

(a)

5-1 (a) (b): A well-organized work site is safe and efficient. Organize materials and equipment so they are close to the walls that need plastering. By aligning functions, such as sifting and mixing, you can save a lot of work.

located close to the mixer to minimize labor. A little extra exertion never hurt anyone, but if you're going to be mixing lots of plaster, a little added effort — say an extra step or two between the mixer and the dirt pile — starts to add up over the course of a few days. Bob Campbell, who lives in Boulder, Colorado, and who has been plastering walls for 17 years, emphasizes the importance of efficiency. "Remember, you're doing something that's incredibly repetitious. Although there is no single thing that will save huge amounts of time, there are twenty things each of which can shave off a half of a second. When combined, they'll save you loads of time!" Bob adds, "Plastering is pretty hard work. You don't need to go out of your way to make it harder than it already is! Although it is going to take you time to learn the skills, you can figure out the mechanics and set up to eliminate wasted effort and decrease work very quickly."

Be Sure There's Plenty of Clean Water Available

Making plaster requires a lot of water, so be sure there is an abundant supply of clean water available. If a well has been drilled and you have access to hose bibs, you can tap into this supply. (Water supply should be established before plastering begins because all of the plumbing should be installed.) If the house you're plastering is to be supplied by catchwater, you may need to fill the cistern before you begin plaster work. Water can be purchased in bulk and is delivered to the site by truck, or you may want to rent a temporary storage tank. Barrels of water may suffice for smaller projects.

(b)

dirt   sifted dirt   mix

Plaster mixing requires lots of water and will also involve some spillage, so be sure the mixing site is well-drained. You don't want to be working in puddles. Gravel, crushed rock, straw, or even non-slip rubber mats can laid down in the mixing area.

Consider Health and Safety at all Times!

When mixing plaster, consider health and safety issues. Clay-rich dirt, sand, pigments, and fiber dust pose a potential health hazard to those who inhale their dust. We recommend wearing a certified dust mask whenever sifting and mixing plasters. Eye protection is also essential: it will safeguard your eyes from inevitable splattering.

Staging also pays huge dividends in safety. A well-organized site is a safe site. Be sure that hoses and buckets — or any other obstacles — are not in the way of work, so you and others won't be tripping over them. Be sure to remove obstructions, both inside and outside. If other workers such as roofers are on the job site, ask them to assist in keeping the site safe and obstruction-free. Keep your tools organized and out of people's way, too. Designate an area for tool storage.

It is a good idea to assign someone on the plaster crew the chore of keeping the site clean and organized. At the end of each work day, ask the crew to clean up. Inspect the site every morning for safety hazards before work begins.

KEEP THE JOB SITE ORGANIZED

Once plastering is underway, the work site can become quite hazardous to workers. Empty buckets, tools, electric cords, and hoses pose a hazard. Make sure you maintain your site during the day while work is in progress. At the end of each day, take a half hour to clean up and reorganize the site. Your efforts will be well worth it!

Provide Ramps and Bridges to Negotiate Obstructions

Remember, you want to be able to deliver full and heavy wheel barrows of plaster to interior and exterior walls without having to negotiate obstructions. Designate the most efficient routes from the plaster-mixing operation to the walls, and make sure that others on the site are aware of them. Clear and level the route.

Sturdy ramps may be needed to pass over door sills. You may also need to provide sturdy bridges over ditches. In two-story homes, you may need to rig up a pulley system to haul plaster to the second story.

Preventing Shock and Electrocution

Make sure electrical cords are away from hoses, buckets, storage tanks, and puddles. Power tools, electric mixers, and other devices pose a serious shock and electrocution danger, especially if you're working with a lot of water, as you will be when making plaster. If cords show exposed wires, replace them. As part of your daily inspection and clean up, be sure to inspect all electrical cords.

Electric lines, along with hoses, sprayers, buckets, and other tools also present a tripping hazard. If other workers are present

5-2: Make sure that the job site is set up so it is easy to transport plaster from the mixing area to the walls in a safe and efficient manner. Obstructions can result in spills and can be dangerous to those transporting heavy loads.

CEDAR ROSE GUELBERTH

on the job site, be sure they're aware of potential dangers. And be certain to run a safety evaluation of your site daily.

Protect Against Gasoline Spills

If you're using a gas-powered mixer, you'll want to be sure that you don't spill gasoline onto the ground or onto plaster components. Spills contaminate the environment and add toxic chemicals to plaster. Avoiding spills is especially important when working on a straw bale home. Loose straw around the site can catch fire that can spread from the loose straw to unplastered walls in a matter of seconds. Even though plastered walls are quite fire resistant, unplastered walls can burn quite readily, incinerating an entire home in short order.

Be especially careful when using a generator or a gas-powered cement mixer — or any other device that poses a fire hazard. Using electric chain saws and weed-eaters for trimming and shaping bales reduces fire hazard. If a spill occurs when filling a generator or a mixer with gasoline, clean the area well and remove contaminated materials immediately to prevent the spread of toxic substances and reduce fire dangers before re-starting the machinery.

Cover Materials and Provide Shade

When they're not in use, plaster components should be covered with tarps to protect them from rain or wind. Large tarps are needed to cover clay-dirt, sand, and straw. Wood pallets and planks can be used to elevate straw bales. Place flour, bagged clays, and pigments on pallets and cover well or store indoors.

Mixing plaster is hot, hard work. To ensure the experience is more pleasurable, you may want to set up your mixing area in a shaded area or provide shade for workers. Cool, fresh drinking water is also a must.

Provide Sturdy, Safe Scaffolding

To reach high sections of walls, make arrangements for good, solid scaffolding. You can rent scaffolding from a local tool rental shop, or make your own out of solid straw bales and planks. Plastering from a ladder is generally not recommended: the chances of falling are too great.

Assemble scaffolding near walls before you begin plastering, and position it for ease of access. Whether you rent scaffolding or make your own, test it first to be certain that it is solid, stable, and safe before anyone is allowed to work on it.

5-3: Mixing plaster is hot, hard work. Make sure the mixing and plastering crews have shade to work in, as shown here, and lots of cool, fresh drinking water.

CEDAR ROSE GUELBERTH

5-4: Steel scaffolding and planks over straw bales help workers plaster the upper reaches of a wall safely.

CEDAR ROSE GUELBERTH

Protecting a House While Plastering

Earthen plaster can make a mess of wood and other finished materials, so mask or tape all finished woodwork and finished materials including window frames, window sills, door frames, posts and beams, tiles, flooring, and railings to protect them from unavoidable splatter.

Protect thresholds and door frames from wheel barrows. Cardboard shields, firmly taped in place, provide good protection. Stairways and floors should be protected, too, although finish work on floors and stairways is best left until after walls are plastered.

What Tools and Equipment Will You Need?

Plaster mixing requires a handful of tools. You'll need sifting screens, shovels, a mortar hoe (a hoe with a hole in the blade), and two sturdy construction wheel barrows (not those cheap plastic ones!) to mix and transport materials. You can also mix the mud in a mixing boat, a cement mixer, or some other container. Making plaster requires hoses,

5-5: This post-and-beam framing wrapped with straw bales will look beautiful when the home is completed, but exposed wood must be masked (as shown here) before plastering to protect the wood.

TOOLS FOR MIXING PLASTERS

- Shovels
- Mortar or standard hoes
- Wheel barrows
- Mixing boat or cement mixer
- Plastic trash cans (preferably Rubbermaid)
- Sifting screens (1/2-inch, 1/8-inch, 1/2-inch, and metal window screen)
- Water source
- Hoses
- Numerous clean five-gallon buckets and an assortment of smaller containers
- Pallets or wooden planks
- Tarps
- Work gloves
- Rubber gloves
- Certified dust masks for all workers
- Safety glasses
- Stiff brush for clean up
- Sponges

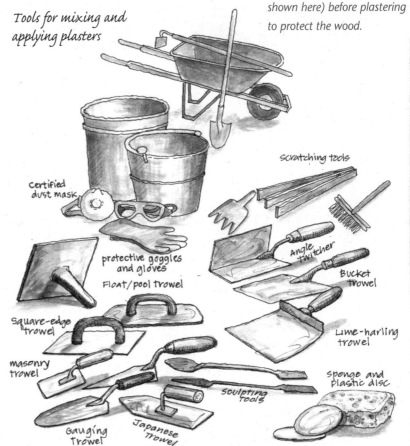

Tools for mixing and applying plasters

Certified dust mask

protective goggles and gloves

Square-edge trowel

masonry trowel

Float/pool trowel

Gauging Trowel

Japanese Trowel

sculpting tools

Scratching tools

Angle Twitcher

Bucket trowel

Lime-harling trowel

sponge and plastic disc

5-6 (a) (b): To screen dirt, (a) a screen attached to a wood frame that fits over a wheelbarrow works well, (b) as do free-standing screens, shown here positioned well — near the mixer and other materials set up for easy access for mixing plaster.

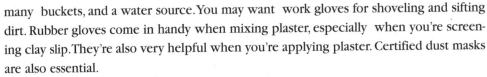

many buckets, and a water source. You may want work gloves for shoveling and sifting dirt. Rubber gloves come in handy when mixing plaster, especially when you're screening clay slip. They're also very helpful when you're applying plaster. Certified dust masks are also essential.

Plastering requires lots of buckets for a variety of reasons — for example, to transport mud to the walls, to clean tools, or simply to hold water for mudders to rinse their hands or trowels. For large projects, five-gallon plastic buckets come in mighty handy; you can pick them up free of charge from many local restaurants, and they're also sold in hardware stores and building supply outlets. Remember to obtain lids, as you may need them to cover plaster mixes overnight. Be sure to provide a variety of smaller buckets and containers as well: containers ranging from one pint to five gallons come in very handy for many different tasks.

JOHNNY WEISS

Sifting screens are required on the job site. They're used to remove rocks and other large chunks such as clumps of clay or roots, from your clay-dirt, which make it difficult to mix and apply plaster. You don't want to be picking them out of your plaster as you're trying to cover a wall. (Clumps of clay remaining after screening and sifting the clay-dirt can be soaked in water and broken down, then used in variety of other steps, for example making clay slip or wall-packing material.)

Several types of screen are used throughout the plastering process, including ⅛-, ¼-, and ½-inch hardware cloth and fine-mesh metal window screen. The simplest screen system for small projects is a section of wire mesh laid over a wheelbarrow, garbage can, or bucket. Dirt is shoveled onto the screen, then sifted. Shaking the screen or running your hands through the dirt will do the job. If you "massage" the dirt by hand till the particles break up, you'll probably want to wear a good sturdy pair of leather or hemp gloves. Cotton work gloves won't hold up very long.

CEDAR ROSE GUELBERTH

A more sophisticated screen consists of mesh attached to a 2" x 2" or 2" x 4" (for a sturdier device) wooden frame. The frame will need to be large enough to fit securely on the top of a wheel barrow (see illus. 5-6 (a)). For larger jobs, you should start with free-standing ½-inch screen. You'll want a screen that is four to five feet high by three to four feet wide, supported by legs at the back. Throw the dirt onto the angled screen, then remove the

SCREENING YOUR DIRT

All dirt needs to be sifted through 1/2-inch hardware cloth to remove unwanted materials. This level of sifting works well for the first two coats of plaster and for infill cob mixes. For finer finishes, use 1/4- to 1/8-inch screen after dirt has been screened through a 1/2-inch screen.

screened dirt from under the device for making plaster. (Note that clumps of clay will fall with the reject rocks. They can be collected and soaked or physically broken down and added to the screened dirt.) After screening dirt with ½-inch mesh, smaller mesh screens mounted on smaller wooden frames placed over wheelbarrow or trash cans are used.

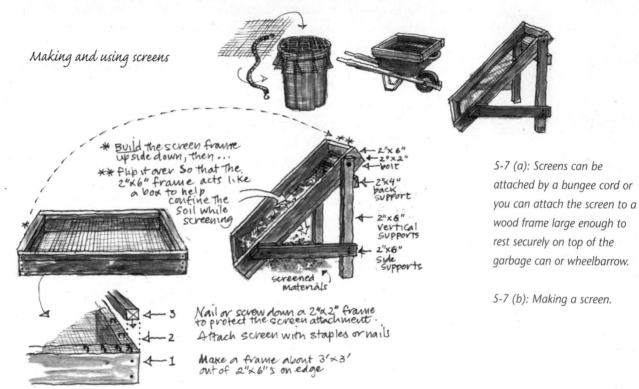

Making and using screens

* Build the screen frame upside down, then...

** flip it over so that the 2"x6" frame acts like a box to help confine the soil while screening

2"x 6"
2"x2"
bolt
2"x4" back support
2"x6" vertical supports
2"x6" side supports
screened materials

3 — Nail or screw down a 2"x2" frame to protect the screen attachment.
2 — Attach screen with staples or nails
1 — Make a frame about 3'x3' out of 2"x6"'s on edge

5-7 (a): Screens can be attached by a bungee cord or you can attach the screen to a wood frame large enough to rest securely on top of the garbage can or wheelbarrow.

5-7 (b): Making a screen.

Good clean-up habits are important on any job site. To do the job right, you will need several good stiff brushes to clean trowels and other equipment. Provide a sufficient number of brushes and sponges so that several people can participate in clean-up. Last but not least, when plastering a wall you will also need some trowels and other tools. We'll talk about them in the next chapter.

Preparing Materials

Before you begin making plaster, be sure that all of your components are ready and that you have sufficient quantities to finish the job. You will need dirt, sand, fiber, and possibly flour paste and manure. (Sand may not be required if there's a sufficient amount in the soil.)

5-8: Be sure you have plenty of materials on hand — including clay-dirt, sand (shown here), and straw — before you begin to make plasters. Many people underestimate the amount of materials they'll need. Assign one person on the crew to keep track of supplies and keep piles well stocked.

Begin by bulk sifting the clay-dirt — that is, run the clay-dirt through ½-inch hardware cloth (or screen) to remove rocks and large debris. For larger jobs, screen the dirt through a standing screen over a tarp or a pit. For smaller ones, you can screen over a wheelbarrow. Cover the material to keep it dry. This dirt is used for the first two coats of plaster. The third coat of plaster, the finish coat, requires finer sifting. Run the dirt through a ½-inch screen, then through a ¼-inch screen. For even finer mixes, you can run it through ⅛-screen.

If your preliminary tests (for example, the jar test) reveal you need to add sand to your plaster mix, we recommend purchasing screened and washed masonry sand from a gravel or sand pit. If you're using another type of sand, be sure to sift your sand through a ¼-inch screen to get rid of the largest stuff, then screen it through ⅛-inch hardware cloth. This sand will be suitable for all plaster mixes. For finer finish coats and alises, you may want to run this sand through a metal window screen — or purchase some finer-grain silica sand.

Next comes the fiber. We explained how to obtain straw fibers and how to chop them up in Chapter 4. Be sure to store all straw in dry areas off the ground and use only clean, dry straw.

You may also need cooked flour paste. We'll describe how to make flour paste in the next chapter. If you're going to be using manure, try to obtain the freshest possible stuff you can. Sift it through ½-inch screen first, rubbing it across the screen to break up the clumps or crushing clumps by hand.

Mixing Plaster

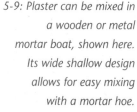

5-9: Plaster can be mixed in a wooden or metal mortar boat, shown here. Its wide shallow design allows for easy mixing with a mortar hoe.

Once your site is set up for plaster production, you are ready to go. Your raw materials are stockpiled in a convenient location, your sifting system is set up, and your tools and equipment are ready — they're all in a convenient location for safe, efficient production.

On large projects, you should place one person in charge of materials — to ensure a steady supply. If someone isn't monitoring materials, you may run out of one component and have to shut the entire production down while you wait for raw materials to be delivered or for more materials to be prepared.

As noted in Chapter 4, one of the keys to successful plaster is mixing your plaster properly. Thorough mixing ensures even distribution of materials and also enhances the binding of clay to other materials in an earthen plaster — so long as there's enough water in the mix. Thus, water and thorough mixing are essential to create a good plaster.

Plaster jobs vary in size and there are many devices to mix plaster, ranging in size and sophistication. Generally, the larger the job the more sophisticated and automated or mechanical the mixing becomes. The reason? To save labor in what is a fairly labor-intensive process.

Mixing Plaster for Small Jobs

The wheelbarrow is commonly used for smaller jobs, but you'll need a deep, sturdy model, not one of those shallow metal or plastic ones. Cheap wheelbarrows only cause headaches. Don't even attempt to mix plaster in one; they're just not strong or stable enough.

Plaster boats can also be used. A plaster or mortar boat is a wooden structure usually made out of ¾-inch HDO (high-density overlay) plywood — the kind of waterproof plywood used for forming concrete foundations — or wood planks. Plaster boats are also made of out sheet metal. Either way, you want a wide, shallow design in which the plaster can be mixed with shovels, mortar hoes, and by hand or foot.

Plaster can also be mixed on tarps or in small pits. A plaster pit can be made by digging out a depression in the ground which is left open or lined with tarp. Far easier to make is a straw bale pit. Straw bales are placed in a square or rectangle that is draped with a waterproof tarp. Be sure to place a piece of plywood or wood planks over the tarp at the bottom of the pit so that shovels used to remove the mix won't puncture the lining. Plastic trash cans are used as well, but not for mixing plaster. They're more appropriate for making clay slip, which you'll need for coating straw bales. (We'll talk more about clay slip shortly.)

Some mudders find that earthen plasters work best if they're allowed to sit for a while after mixing — from one-half to four hours. This allows the clay and straw to become more pliable and increases the binding capacity of the clay, which means that the clay will do a better job of binding to the sand and straw. On larger projects or when you're on a tight work schedule, it may not be practical to wait and it is not necessary.

Mixing Plaster for Large Jobs

For larger jobs, good-sized pits come in handy. Cement mixers also work well for larger plaster jobs; gasoline and electric models are available. (Electric mixers are a cleaner option —

5-10: Plaster can be mixed by foot in a pit excavated in the ground or in a straw bale pit lined with a waterproof tarp (as shown here). Plywood or wood planks on the bottom of the pit (over the tarp) allow for easy shoveling and prevent the tarp from being punctured. This method is not only enjoyable, it permits one to make a lot of plaster quickly.

5-11: Cement mixers with paddle-like blades attached to a rotating barrel like the one shown here work best for mixing plasters. Mortar mixers with stationary barrels and rotating tines can cause problems as straw wraps around the tines and gums up the works.

no exhaust or gasoline spills!) Cement mixers come in many sizes and can be rented by the week. For large projects, it is often cost effective to buy a unit, then sell it after your project is complete.

Don't rent a mortar mixer. In mortar mixers, the barrel is stationary and the tines rotate inside it. Straw wraps around the blades and gums up the works. You'll spend more time cleaning the device than mixing! Cement mixers on the other hand have paddle-like blades. The blades are attached to the barrel, which rotates, allowing the material to mix more freely.

As a final note, when setting up the mixing area, be certain that you can maneuver the wheelbarrow in place and dump the plaster into it, then send it on its way.

Team 1
Monitor & prep materials

Team 2
Mixing & delivering to walls

Team 3
Plasterers

5-12: Working in teams, each with a specific function, helps to increase the speed and efficiency with which plaster is made and applied.

Working in Teams

For large jobs, it is good to designate teams. One team preps materials and another mixes plaster and delivers the mix to the walls. Then another team applies the mix. The number of people on each team depends on the size the job.

Working in teams divides labor and responsibilities up, resulting in a smooth-running operation. You may want to switch jobs to reduce repetitive strain injury and boredom. However, people often find that they prefer to stick to one task — one they're particularly skilled at doing.

How Do You Mix Plaster?

One of the keys to mixing plaster is to evenly distribute the materials while mixing. If you're using a cement mixer, start with about one-third of the water you expect to use, then add some sand. Let it mix, then add clay-dirt. If your plaster recipe calls for ten parts sand and ten parts clay dirt, for instance, begin by adding five shovelfuls of sand while the mixer is running, then add five shovelfuls of dirt. After that's had a chance to mix for a couple of minutes, add some more water and then throw in the remainder of the sand and dirt, giving each a little time to mix in. If the mix becomes dry, add water.

You'll know your mix is too dry if the components don't mix well. They also tend to cake onto the walls of the mixer. These are indicators that you may need to add more water. Proceed cautiously, however. It is easy to add too much water. There's a fine line between too dry and too wet and it is very easy to cross that invisible line!

Once the dirt and sand have mixed, it is time to add flour paste. Manure can be added anytime. Fiber generally goes in last. Note that straw and some other forms of fiber absorb moisture, so you may need to add a little water at this stage. But what happens if you do add too much water to a plaster mix? The easiest way to remedy the problem is to add more of the dry ingredients in proper ratios until you achieve the desired consistency.

Mixing plaster takes a little experience. You need to know the final consistency you're shooting for; add water slowly, observing and feeling the plaster as you proceed. Different coats require different consistencies, a subject we'll explore in Chapter 6.

As a final note on the subject, be sure to ask for feedback from your plaster crew. They'll let you know if the mix is too wet or too dry and how it is going onto the wall. Adjust your mix according to their comments.

Some Additional Pointers.

No matter whether you are mixing in a wheelbarrow, a pit, or a cement mixer, add ingredients slowly and evenly. Avoid the temptation to add all of your materials at once; it is much harder to mix thoroughly if you do. In a mixer, sand or dirt may compact on the sides of the mixer and the barrel will rotate without mixing. If this occurs, you may need to add more water or stop and scrape the material off the sides of the barrel.

In a wheelbarrow, pit, or mortar boat, sand or dirt may become compacted along the bottom or on the sides if added all at once; this is very difficult to break up. Even when mixing in a boat or pit, be sure to put half of the sand in first, then add half of the dirt with water. Mix, then add the remainder of the material. If you don't, you'll have trouble mixing the stuff.

Finally, always seek feedback from your mudders and adjust the mix accordingly.

Watch Out for these Common Errors

Mixing plaster takes a little practice, but success comes quickly. Watch out for these common mistakes: adding too much water initially; dumping all the ingredients in at once; putting the dirt in first; adding too much water as you go along; and losing count as you are shoveling in components.

Dry-Mixing Earthen Plaster

Earthen plaster can also be dry-mixed. First, mix all of the dry ingredients. Then add wet ingredients, such as water, flour paste, and cactus juice — a little at a time, stirring the mix so components are evenly distributed. Too much water at one time can

5-13: To dry mix plasters, begin by adding all of the components, mix thoroughly, then gradually add water. Continue mixing, adding as much water as necessary to achieve desired consistency. Be sure to mix from the bottom for thorough mixing.

CEDAR ROSE GUELBERTH

cause the dry materials to stick to the sides of the mixer or the bottom of the wheel barrow, mortar boat, or mixing pit.

A Note on Measuring Materials

Most plasterers add components to a mix by the shovelful, counting as they go. When measuring this way, it is important that the shovelfuls are of equal size for all materials. It is also important that one person do the shoveling to maintain consistency. If not, ratios can change and the plaster mix may be seriously altered. A more accurate way of mixing components is to use one-to-five-gallon buckets. Filled to the same line each time, they leave little room for error. It is a good idea to have an extra person to count, especially if you tend to get confused or distracted and lose track of the number of shovelfuls or bucketfuls of dirt or sand you've added. Ask an assistant to help keep track, too.

Plaster Mixers Extraordinare!

In most work crews, one or two people emerge as the expert mixers. They seem to have a knack for this sort of thing and even enjoy mixing batch after batch of high-quality earthen plaster. A good mixer not only produces a good material to work with, he or she also produces it at a constant rate, always one step ahead of the mudders.

Plastering Overview: How Many Layers and What are they Called?

One of the most confusing aspects of earthen plastering to newcomers has to do with the coats — more specifically, how many coats of plaster are required and what each is called. We can bring some clarity to the issue.

Why Apply Multiple Coats?

Gernot Minke notes that two or three layers of plaster should be applied in most situations. Why is plaster applied in coats? Why not just slap a thick coat on and be done with it?

A layered system — as opposed to a single, thick layer of plaster — works best for a number of reasons. Three thinner layers allows one to build a thick, protective layer. It is less likely to crack than a thicker layer applied in one coat. And if cracks develop in a layer, they won't go all the way through the plaster, as they would if a single, thick layer were applied. In addition, a single, thick layer of plaster — virtually any plaster — may slump or bulge and thus come loose from a wall during or just after application.

An Exception to Every Rule. There is at least one exception to the multiple coat rule. Bill and Athena Steen and collaborators in Mexico have come up with a straw-clay plaster which works well in their arid climate. With the local soils, they use a little less than one part sand, two parts clay soil, and three parts chopped straw. This straw-rich plaster goes on very thick — up to two inches — without cracking or slumping. Bill and Athena have applied it as a single coat with good results. They've also applied finished coats over the plaster. This system is effective in arid climates. (We'll discuss this system in more detail in Chapter 6.)

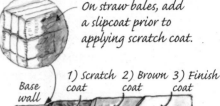

On straw bales, add a slipcoat prior to applying scratch coat.

Most Plasters Require Three Coats

As explained in Chapter 2, most mudders apply three layers of plaster on straw bale walls: (1) a scratch coat, (2) a brown coat, and (3) finish coat. This system produces a stable, long-lasting, and durable finish in all climates. As you shall soon see, straw bale walls also require the application of a coat of clay slip over the straw before plastering, to key the base coat into the surface of the bales. Earthen walls, in contrast, usually require only two coats of plaster: (1) a base coat and (2) a finish coat, a topic we'll discuss further in Chapter 6.

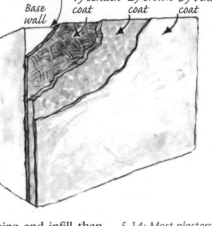

1) Scratch coat 2) Brown coat 3) Finish coat

Base wall

The reason earthen walls require only two coat is two-fold. First, they're made from clay-rich dirt, the same material used to make earthen plasters. Plaster readily adheres to like surfaces. Second, earthen walls are generally smoother and require less shaping and infill than straw bale walls. As a result, they require less plaster build-up to achieve a smooth, flat surface, although there are some notable exceptions, such as rammed earth tire walls.

One of the reasons newcomers are confused about plasters at the outset is that different plasterers name the layers differently (see side bar). The first coat of plaster, the base coat, is most commonly called the *scratch coat*. Cedar coined the term *discovery coat* for straw bale work, because this coat of plaster reveals the topography — the peaks and valleys in a straw bale wall. It allows the mudder to discover the topography of the wall.

The second coat of plaster is known as the *brown coat*. Cedar likes to call it the *infill coat* because it helps to fill in and smooth the surface.

The final coat is known as the *finish coat* or *color coat*. It is often troweled smooth and is often pigmented, giving the wall color. As you will soon see, the finish coat can be left unpigmented or can be painted with a variety of natural pigmented finishes, such as an alis (clay "paint"), or finished with clear oils.

It really doesn't matter what you call the various coats. The important thing is to understand how each coat is prepared and applied — and what each one does. We'll shed some light on this shortly.

5-14: Most plasters are applied in three coats — the scratch coat, the brown coat, and the finish coat (shown here). In straw bale construction, a slip coat facilitates the application of plaster and results in a better product. For earthen buildings, a base coat and finish coat are often sufficient.

MULTIPLE COATS/MULTIPLE NAMES

- First coat — aka base coat, scratch coat, or discovery coat
- Second coat — aka brown coat or infill coat
- Third coat — aka finish coat, color coat

Before Plastering

Before plaster can be applied to a straw bale home, the walls must be properly prepared. That means that walls should be plumb. They should be shaped and shaved with a weed eater, electric chain saw, or some other device to remove loose straw (fluff) which plaster will not key into well. Walls should be stable and pinned. In the case of load-bearing walls, the top plate should be securely attached to the foundation and bales should be compressed, so they won't settle and crack the plaster (Chapter 3).

As you may recall from Chapter 3, compressing load-bearing walls occurs naturally after the roof is installed. This takes about six to eight weeks. Walls can also be artificially compressed by using straps or cable that extend between the top plate and the foundation or all-thread, which runs from the foundation through the wall to the top plate. By tightening straps, cables, or all-thread, a builder can achieve compression in a few days that a roof will take six to eight weeks to accomplish.

In nonload-bearing structures, plastering can begin as soon as the walls are erected — that is, so long as you've built your walls with good-quality (dry, compact) straw bales. Once bales are secured and prepared, plastering can begin immediately because there should be little, if any, settling. In nonload-bearing structures it is the post-and-beam, modified post-and-beam, or framed structure built from wood or other materials that supports the roof and the bales. Because the load is being carried by the framing members, the bales will not be compressed by the roof.

For more advice on settling and compaction of walls, please review the material in Chapter 3 on wall building. You may want to consult a knowledgeable straw bale builder if you have questions on this procedure. Many books on straw bale building also cover the subject, and in a lot more detail. *The Last Straw* also routinely publishes new ideas on compressing load-bearing straw bale walls. Taking a workshop is also extremely important. These and other sources of information are listed in the Resource Guide at the end of the book.

Install all Plumbing and Electrical

Before plastering begins, be sure that electrical wire and plumbing are installed. On interior wall surfaces, boards or other devices required to anchor shelves and cabinets to

---

IN A HURRY?

If you are in a hurry to finish a load-bearing straw bale home, you can apply a protective slip coat (we'll explain why shortly) and a scratch coat to protect your straw bales from the elements before the roof has compressed the walls. Settling will cause these coats to crack, but that is okay, you'll take care of the cracks in the brown coat and finish coat, which shouldn't be applied until the walls have completely settled.

GOOD BALES MAKE GREAT WALLS

A good straw bale for building is tightly compacted, held together with polystring (not wire), golden yellow, clean, dry, and free of mold inside and out!

straw bale walls should be placed in the wall before plastering commences. Interior walls should be built, too.

Leave the Chicken Wire in the Barn

Although chicken wire and metal lath are often applied to straw bale homes coated with cement stucco, they're totally inappropriate when earthen plasters are being used. They don't function as well as some would think. Earthen plasters adhere better to straw, creating a strong bond between bale and plaster. Chicken wire and lath, in fact, will work against you by preventing the adhesion of mud plaster to the bales. Air pockets form beneath the plaster and these areas will eventually deteriorate and crumble off.

Keying Layers

With earth plasters, it is important to key each layer into the previous one. That means that each layer must bond well to the one before it. The finish coat, for instance, is keyed into the infill or brown coat beneath it. The infill coat is keyed into the discovery or scratch coat. The discovery coat is keyed into the straw via the slip coat.

In straw bale structures, Cedar Rose, working with Matts Myhrman and others, pioneered a technique designed to ensure the tight adhesion of the base coat to the underlying straw. Using a drywall texture gun or pneumatic sprayer, they spray the straw with a thin layer of clay-dirt, known as *clay slip*. It penetrates the surface of the bales and thoroughly coats all exposed straw, thus forming a tight bond with the scratch coat.

Steps Involved in Plastering a Straw Bale Wall

With these basics in mind, we turn to the main plastering system we're going to be teaching you. This system begins with two prep coats, the adhesion coat and slip coat.

The Adhesion Coat

Before you can begin plastering, wood and metal that will be covered with plaster can be coated with an adhesion coat, a material to which the plaster better adheres. Use a mixture of manure, cooked flour paste, and sand and apply it to exposed wood by hand or with a paint brush. This application is called an *adhesion coat*. An adhesion coat seals porous surfaces and reduces the absorption of moisture from newly applied plasters and provides "tooth" to smooth surfaces for plasters to key into. It can also be used to ensure a tighter bond

5-15: Workers apply clay slip using a drywall texture gun. Applying a coat of clay slip to straw bale walls provides a solid base to which plaster easily adheres.

5-16: An adhesion coat applied over wood, metal, concrete, and nonporous materials can provide a rough, adhesive surface for plasters to adhere to.

between earthen plaster and concrete, polypropylene earthbags, plywood, and tires. (Application of adhesion coat is not necessary on spans of four inches or less.)

On the exterior, an adhesion layer is suitable for smaller areas — for example, an 8-inch wood frame. Don't expect it to work on a 4- x 8-foot sheet of oriented strand board (OSB) sheathing that's to be coated with plaster — for example, on walls on the gable ends of a house. Plaster over large expanses of wood, such as OSB or other sheathing materials, or concrete won't stand up to weather. Over time, the plaster will be eaten away by the weather. For such areas, you should consider another type of design — for example, installing wood siding over a large expanse of sheathing, rather than plaster. Or design your home in such a way as to minimize or eliminate large expanses of wood or concrete which are to be coated with plaster.

The adhesion coat works well for covering smooth interior surfaces — large or small. Interior walls are protected from temperature extremes and the adhesion coat provides sufficient anchoring for plaster so long as the walls are not exposed to humidity for long periods, as in bathrooms.

Once the adhesion coat is in applied, it is time to apply the slip coat to the bales.

The Slip Coat

The slip coat is a layer of clay slip, approximately the consistency of heavy cream or very thick chocolate milk, which is applied to straw bale walls by machine. The slip can, for instance, be sprayed on using a drywall texture gun, stucco pump, or pneumatic sprayer. Stucco machines or hand pumps also work.

The slip coat penetrates the surface of the straw bale wall and coats the straw with thin layer of clay-dirt. This provides a solid, like surface for the next plaster layer, the scratch coat, to bond to or key into. As noted in Chapter 3 and described more fully in Chapter 6, after the clay slip is applied and is still wet, crevices in the wall between bales are stuffed with dry straw and then packed with straw soaked in clay slip. This helps create a solid base for plasters and an air-tight wall.

The slip coat is applied only to straw bales. Window bucks (rough window frames) or framing members to which plaster will be applied are coated with the adhesion coat and thus are not sprayed with clay slip. We'll tell you how to make clay slip in the next chapter.

5-17: Packing the holes and crevices in straw bale walls with a straw/clay slip material eliminates air pockets and cold seams and provides a base for plaster to adhere to.

APRIL ROSE GUELBERTH

The Scratch Coat (Discovery Coat)

The first layer of plaster is a wet, sticky mix called the *discovery coat, base coat* or *scratch coat*. It is applied by hand. Hand application not only works well, it is lots of fun. In addition, applying by hand provides the control needed to key a plaster into a wall.

The discovery coat consists of clay-rich soil, sand, and fiber. Manure and/or flour paste are also often added, depending on the quality of the clay-dirt. The discovery coat contains a high proportion of clay that helps it key into the slip coat already applied to the straw bales. As a general rule, the base coat is fairly thin, about ¼- to ½-inch thick. It goes on quickly and easily.

The Brown Coat (Infill Coat)

The next coat is the brown coat or infill coat. It is made from a thicker plaster mix with a little more sand and straw than the discovery coat. (That is, it is not as wet and as sticky as the discovery coat.) The mix provides more structure to build out the surface, that is, to fill in holes and to shape and smooth the walls. Cedar calls the brown coat the *infill coat* — a name that aptly describes its function. This coat ranges in thickness from one to six inches (more in some cases).

Depending on your preference, the infill coat can be used to convert the rather bumpy straw bale wall into a relatively smooth, flat plane or a smooth, undulating surface.

SCRATCH COAT

The term "scratch coat" comes from the practice of scratching the base coat after it has been applied. This creates a furrowed surface that keys into the next layer. Scratching the base coat is common practice in cement stucco, but is not typically employed with earthen plasters because the surface of the scratch coat, when applied by hand, is pretty rough. The next coat of plaster, the brown coat, adheres to the scratch coat well.

5-18: The scratch coat, also known as the discovery coat, creates a solid base that the next layer of plaster keys into. In straw bale construction, it also reveals the topography of the wall.

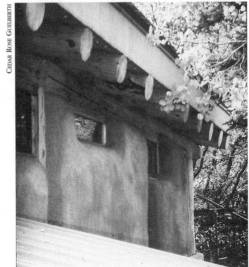

5-19: The infill coat, applied over the discovery coat on a straw bale wall increases the thickness of the plaster, providing further protection, and fills in the irregularities in the wall surface, evening it out.

The final result is all up to you — what you want in a wall and how much work you want to put into it. To achieve smooth, flat walls, you'll have to apply enough plaster to eliminate all of the peaks and valleys in the straw bale wall (that means you will have at least a one- to two-inch coat covering the surface). An undulating wall requires less infill plaster. Although the infill coat is used to even and shape walls, it can also be used to sculpt walls — that is, to create unique, artistic surface features.

The Finish Coat (Color Coat)

The last coat gives the wall its final appearance and the desired level of smoothness. It is appropriately called the *finish coat*. Like the other coats, it contains clay-dirt, sand, and fiber. In this case, however, the clay-dirt and sand are typically more finely sifted.

For the finish coat, we use more clay and less sand than underlying layers. This helps us to achieve a sealing effect. Finer sand can be used, as well. Manure, smaller strands of straw, wool, hemp, or cattail fluff may also be added to the finish coat as fiber.

Besides defining the texture, the finish coat also provides color. Mudders, therefore, often refer to it as the *color coat*. As you will learn in Chapter 7, finish coats may also serve as a base for pigmented color finishes, including alises, decorative clay finish coats, lime wash, silica paint, or milk-based (casein) paints. Clay paints can be manipulated to change the texture of the wall, too. Clay paints are also known as *alises*. Traditionally, alises were applied as a clay slip over plastered walls, providing protection and color.

The finish coat should not dust, and should form a hard, consistent, durable, and attractive surface. Flour paste, cactus juice, powdered milk, casein or other additives are often critical to the success of a finish coat.

Likes onto Likes, But Wet Your Walls First

All layers of plaster are applied to a like surface for best results, as just noted, and each surface is wetted before the next layer is applied.

Wetting surfaces is important because it creates a strong bond between each layer. When a wet second coat is applied to a dry first coat, water is wicked out of the new plaster by the dry, underlying material. This causes the new layer to dry quickly at the interface. The result? A crumbly, weak attachment zone. Consequently, the new layer of plaster does not adhere well to the old. So don't forget to wet surfaces to be plastered.

WET OR DRY?

Apply slip coats and adhesion coats to dry surfaces. Apply the discovery, infill, and finish coats to moistened surfaces for maximum bonding between coats.

Quick Recap

In this chapter you've learned about creating a safe, efficient worksite, the tools and materials you will need, and how to mix these materials to make plaster. When plastering a straw bale home, you begin by applying an adhesion coat on wood and non-straw components over which you want to apply plaster. The adhesion coat binds the first layer of plaster (the discovery or scratch coat) to the underlying substrate, such as wood, metal, or concrete. Before the first coat of plaster is applied to a straw bale wall, we also recommend spraying a layer of clay slip on straw bales. This layer, known as a *slip coat*, helps the discovery coat key into or bond with the straw bales to create a strong, stable base for subsequent layers of plaster. We refer to slip coat and adhesion coat as the *prep coats* — for they prepare the surfaces for plastering. After the slip is applied to the straw bales, inside and out, the walls are stuffed and packed. Then comes the plaster.

Earthen plaster on straw bale homes generally goes on in three coats. The first is the *base* or *scratch coat*. We like to call it the *discovery coat*, for it reveals the irregular surface topography of the straw bale wall. The first plaster coat must key into the straw bale wall. The second coat is the *brown coat*, which we call the *infill coat*, a term that aptly describes its function, which is to fill in the low spots and to shape the wall surface. The infill coat creates a relatively smooth surface. The third and final coat of plaster is known as the *finish coat* or *color coat.*

Three coats of plaster typically provide a one- to two-inch protective layer over an entire wall. Bear in mind, however, that some areas may be thicker than others. Low points may need to be built out to three to six inches (15 cm) thick, while high points may be 3/4-inch thick, the minimum recommended thickness. (In other words, you don't want your plaster any thinner than 3/4 inch.) Although interior and exterior straw bales walls are exposed to dramatically different conditions, they both receive three coats of plaster.

With this information, we can now move on to our next topic: applying plaster.

CHAPTER 6

Applying Earthen Plaster to Straw Bale Walls

S UCCESSFUL PLASTER JOBS require attention to details in the design and construction of a home. They also require a well-made plaster — that is, a plaster that contains good quality materials in the right ratios. Plasters must also be thoroughly mixed and applied properly. If made correctly and applied well, an earthen plaster will go on a wall easily and provide maximum protection with minimal maintenance and repair.

Plaster application requires knowledge, skill, and practice. In this chapter, we'll provide a solid foundation for understanding plaster mixes and techniques for applying plaster. Let's begin by examining the tools of the trade.

Plastering Equipment

Besides the tools required to mix plaster, you'll need an assortment of tools to apply plaster to walls — as well as some safety gear. Certified dust masks are needed when sifting materials, mixing plasters, and working with powdered clay and various pigments — even natural earth pigments. Eyeglasses or goggles are recommended to protect the eyes of those who are mixing plaster and applying it to the walls, during which splattering can occur. Masks and protective eye wear are critical for spray applications.

Quality work gloves are useful, too. Although earthen plasters are gentler on the skin than lime plaster or cement stucco, the sand can be rough on your hands. Clay can dry hands, too. We recommend rubber gloves for mixing and applying plaster, and regular work gloves for sifting and shoveling dirt.

TOOLS FOR APPLYING PLASTER

- Rubber gloves
- Certified dust masks
- Protective eyewear
- Trowels
- Sponges and clean rags
- Buckets (lots of five-gallon and several smaller sizes)
- Two wheelbarrows
- Hose with spray nozzle
- Buffing disks (optional)
- Compressor and drywall texture gun for slip coat (optional)
- Stucco spray machine (for machine application of plaster) (optional)
- Hawks (optional)
- Mortar boards (optional)

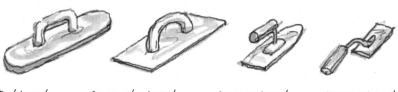

Pool trowel   Square edge trowel   Japanese trowel   Masonry trowel

6-1:There are many types of trowels to choose from. A few of the most frequently used are shown here, including the common pool trowel with rounded edges for easy plaster application and the Japanese trowel which is light and fits easily into your hand.

6-2: Plaster can be transferred from wheelbarrows or buckets to mortar boards or buckets set up near the wall. Set the mortar boards on a stable base that is high enough for ease of access by several mudders.

Trowels are essential to plastering. They come in several types and sizes. Most mudders use steel trowels. (Figure 6-1) Those with square edges are difficult for beginners. Rounded-edge trowels, known as pool trowels, however, work quite well for beginning mudders: their rounded edges make them less likely to slice into walls. Japanese trowels are an excellent choice: they're made of flexible stainless steel or hard steel, and the small handles fit comfortably in most people's hands. Sponges are useful for buffing walls. Yogurt container lids can also be used to buff finished walls, but be sure to trim off the edge (or lip) of the lid and shave off any dimples at the centers of the lids.

Plaster can be applied right out of the wheelbarrow, but in many cases mudders transfer plaster into buckets or onto a mortar board next to the wall on which one is working. A mortar board is a flat piece of plywood or a wooden platform made from planks, usually no more than three feet by three feet (1 m by 1 m), placed on a stable base — for example, on top of several upturned five-gallon plastic buckets, a straw bale, or sawhorses. (Figure 6-2) A mortar board holds a lot of plaster and can be accessed by several people, though for larger crews, you'll need several mortar boards.

Mortar boards and buckets reduce the number of trips you will need to make to the wheelbarrow, and are especially helpful when you've got several people working on a wall. The mortar board helps the plaster-mixing crew stay ahead of the mudders — that is, they can empty their wheelbarrows and return to the mixing area more quickly. Be sure to elevate mortar boards and buckets to reduce back strain of mudders caused by bending over to scoop up mud all day.

Be sure to have plenty of buckets, sponges, and brushes for clean up. Sponges are useful for cleaning off trim and other surfaces.

The Prep Coats

Before the first coat of plaster is applied to a straw bale wall, we recommend application of an adhesion coat and slip coat, which we refer to as the *prep coats*. Let's begin with the adhesion coat.

Adhesion Coat

Before you begin plastering, you need to apply an adhesion coat to wood, metal, and concrete on which plaster will be applied. The adhesion coat helps bind plaster to these materials and consists of cooked flour paste and manure mixed with gritty sand.

PREPARING THE ADHESION COAT. The adhesion coat is made from two parts cooked flour paste (see recipe on page 114), one part fresh screened manure, and one part coarse sand. These ingredients are mixed in a bucket to form a paste that is applied in a thin layer onto wood and other materials over which plaster will be applied.

Don't expect miracles from an adhesion coat, however. On interior surfaces, it can be applied successfully on larger spans of concrete, drywall, or sheathing, but on exterior surfaces it doesn't perform well on spans of cement or sheathing larger than two feet (60 cm) across unless they're well-protected, as noted in Chapter 3.

APPLYING THE ADHESION COAT. The adhesion coat can be applied by hand or with an ordinary paint brush. A four to six-inch brush works well. (Figure 6-3) Spread the adhesion coat evenly over the wood or other surfaces and be sure to stir the mix continually as you apply it, to prevent the sand from settling to the bottom of the bucket. One coat is sufficient, although you can apply a second coat if the first coat isn't gritty enough. Apply enough material to the surface to achieve a good gritty layer of sand about $^1/_{16}$- to $^1/_8$-inch thick. Let the adhesion layer dry completely before you apply the first coat of plaster (the discovery coat).

Slip Coat

Next comes the slip coat — a layer of clay-rich dirt mixed with water — which is applied wet to straw bale walls, and penetrates and coats the surface of the bales, keying into them to create a secure base to which the discovery or scratch coat will easily adhere. (Note: although clay slip needs to have a high clay content, it is not

> ## TOOLS AND MATERIALS FOR MAKING AND APPLYING ADHESION COAT
>
> - Cooked flour paste
> - Manure sifted through 1/2-inch screen
> - Gritty sand
> - Bucket
> - Paintbrush (optional — usually applied by hand)

APRIL ROSE GUELBERTH

6-3: A straw bale wall with a wood post that will be covered with an earthen plaster. Note that the post has already been coated with the adhesion coat. Once dry, the adhesion coat (a mixture of cooked flour paste, manure, and sand) creates a rough surface to which the subsequent layer of plaster adheres easily. The adhesion coat also minimizes absorption of water from plaster by the underlying materials, which would result in a weaker bond. Here, the scratch (discovery) coat is being applied by hand.

MATERIALS AND TOOLS FOR MAKING AND APPLYING SLIP COAT

- Prescreened dirt (dirt run through 1/2-inch screen)
- Cooked flour paste
- 30-gallon Rubbermaid* trash cans or equivalent (with lids)
- Shovels
- Metal window screen
- Bungee cord
- Either cement mixer or trash can with two hoes for mixing
- Eight or more buckets
- Drywall texture gun or other sprayer (e.g., stucco gun with correct nozzles)
- Compressor

*Recommended because they are a good size, flexible and stable, won't crack, and have handles on the top and bottom.

6-4: Clay-slip can be sprayed onto straw bale walls using a drywall texture gun which ensures that it penetrates the surface of the bales.

APRIL ROSE GUELBERTH

pure clay. If it has a low clay content, you can add flour paste to make it bind better.)

Clay slip may be applied in one of several ways. It is often sprayed on with a drywall texture gun. (Photo 6-3) Be sure to apply this coat very carefully, so that it thoroughly coats the straw and penetrates the surfaces of the straw bales. When the clay slip is applied correctly, you won't see any yellow from the straw: the walls should be completely coated with clay-slip.

The slip coat bonds with the straw, providing a like surface to which the first coat of plaster (the discovery coat) adheres. Remember, when plastering always apply like materials to like materials. That way, you'll be able to achieve a solid base and a strong bond between the bales and the plaster.

MAKING CLAY SLIP. Clay slip is made by extracting clay from subsoil. When clay-dirt is mixed with water, the clay particles and silt become suspended in solution, while sand and aggregate fall to the bottom. The solution — called *clay slip* — is screened to remove aggregate, sand, and debris and then sprayed on the walls. This process is simple and fun — and it is a great way to begin learning about plaster materials.

Here's how you make clay slip from dirt using a plastic trash can. First, sift your soil through a ½-inch screen to remove rocks and organic debris. You can sift onto a tarp or into a wheelbarrow or other container, breaking up any small clumps of clay with your hands by rubbing them against the screen. Next, fill a clean plastic trash can one-third full of water. Shovel the sifted subsoil into the trash can while stirring with a hoe (See Chapter 4, Figure 4-6). As you stir the mix, clay becomes suspended in the water. (Run your hands through it: it really feels cool.) Sand and aggregate (small pebbles or rocks) settle to the bottom as you stir and will continue to settle out after you stop stirring.

When the clay-water mixture is as thick as heavy cream, it is time to start "harvesting" clay slip. Scoop the clay-water material using a bucket or some other container, being sure to avoid the gritty dregs at the bottom of the garbage can.

If you are going to be using a drywall texture gun (or hand sprayer) to apply the slip coat, the material must be poured onto a metal window screen (plastic window screens rip too easily) placed over a second clean trash can or another container. The screen can be attached with a bungee cord or secured to a wooden frame that fits tightly over the top of the garbage can (see Chapter 5). Fine screening of the clay slip removes large grains of sand, debris, and residual aggregate, which can clog the tiny openings in the spray nozzle. You'll very likely have to work the slip with your hands or a trowel. (Figure 6-5) Rubbing the mix over the surface of the screen not only helps work the clay through, it speeds up the process considerably. Rubber gloves may be worn to protect your hands. (If the mix is so thick that you're having difficulty working it through the screen, add more water to the slip mixture. Don't add too much, however.)

You'll have to clean the screen from time to time to remove the sand and stones that accumulate on the surface. If you're using a framed screen, turn the screen upside down and rap it with your hands to remove the sand that accumulates on its surface, or scrape it clean with a trowel. Screens that are attached to the can via bungee cords can also be scraped clean with a trowel or a board.

While screening the mix, you can add cooked flour paste. We recommend anywhere from two cups to two quarts of flour paste for a 32-gallon trash can full of slip, depending on the clay content of the subsoil you are using — the less clay, the more flour paste.

As you pour each bucket through the screen, clay slip begins to accumulate in the trash can (or bucket or tub) under your screen. Check it to be sure that it is not too watery: it needs to be like heavy cream. If it is too watery, add more clay-dirt to the first mix and stir with your hoe. This will put more clay and silt into the solution.

> ## USING A CEMENT MIXER TO MAKE CLAY SLIP
>
> If you want to, you can use a cement mixer to make clay slip. Add some water to a cement mixer, then add some sifted dirt. Let it mix, add more dirt until the mix achieves the consistency of a heavy cream, then pour the mix into a trash can, wheelbarrow, or bucket. Making clay slip in a mixer is easy, but be sure to let it mix a while to liberate the clay particles from the dirt, sand, and aggregate.

6-5: When screening clay slip, work the slip through a fine screen by rubbing the mixture with your hands (wear rubber gloves!) or a trowel. If it's too thick to go through the mesh, add a little water.

TEAMWORK SAVES TIME AND MONEY

For a large project (a whole house, for instance), you may want four teams: one team of two people to sift and prepare materials; another team of two to make clay slip and wall-packing materials; a team of two to transfer the slip to the hopper and spray it on the walls; and a team of four packing walls. Fewer people are required for smaller projects or when individuals have lots of experience.

APPLYING CLAY SLIP. Applying clay slip with a drywall texture gun is a fast, convenient, and easy way to cover walls, providing great coverage and good penetration and creating a great surface into which the discovery (scratch) coat keys easily and well. However, you'll need an air compressor to power the gun and some source of energy — either gasoline or electricity — to run the compressor motor.

You can rent a drywall texture gun and compressor for about US $15 to $25 per day, or you can purchase one. A drywall texture gun with a hopper costs US $80 to $400; a compressor will cost several hundred dollars. Or, you can purchase a drywall texture gun and rent a compressor. Whether renting or purchasing a compressor, be sure to get a unit that provides enough pressure to run a texture gun consistently: a four-horsepower, 13-gallon capacity unit works well.

Yet another option for applying the slip coat is a pneumatic pump sprayer, such as a Quick Spray or a stucco gun. Basically, any applicator with a high-pressure spray capability will work, although you may need to purchase a special nozzle for applying clay slip. Some units are sold with several different size tips for different applications.

Pneumatic pump sprayers and stucco guns provide great coverage and superior (deep) penetration. Unless you're a professional who will be plastering a lot of houses, however, the price of this equipment can be prohibitive. Some owner-builders purchase a unit for their project, then sell it once the house is complete. Suppliers are listed in the Resource Guide.

Clay slip can also be sprayed on the wall with a hand-powered pump applicator. A hand-pump garden sprayer with a short hose fitted with a larger nozzle is simple and inexpensive. Pour in the clay slip, pump it up, then spray the stuff on the wall. While this device provides good coverage, the slip does not penetrate very well; in other words, this technique is much less effective than other methods we've described. We consider hand-powered applications as a last-resort option, but they are acceptable if a compressor is not available.

When applying the slip coat using a drywall texture gun, pneumatic pump sprayer, or some similar device, be sure to apply it evenly. Be certain to overlap "paths" so as not to leave any surface uncovered. Be sure that you spray all areas and that the slip coat penetrates deeply throughout. However, don't apply the slip too heavily. You will know if you are spraying on too much because the slip will drip or trickle down the bales. If the clay slip is too thin or if you have sprayed across the surface too quickly, yellow straw will be showing through. Be sure to go over the wall a second time to achieve good coverage. Finally, be sure not to spray over the adhesion coat applied to wood or other surfaces. The slip coat dries rapidly, converting your beautiful golden straw bale wall into a dirt-colored layer nearly ready for plastering.

STUFFING CRACKS, SEAMS, AND OTHER OPENINGS. Before you can begin plastering and while your walls are still wet, there's one more important step: you need to fill all gaps and crevices in the straw bale wall.

As noted in Chapter 3, straw bale walls are assembled in a running bond pattern (see Figure 6-6). Where bales come together, gaps occur — some large holes and some minor crevices. Check for them by running your hands along the lengths of the bales, pushing them into the seam. Minor crevices uncovered by this simple technique can be filled later with your infill coat.

Your priority in this step is to pack the larger crevices and gaps. As shown in the accompanying illustration, most often gaps are found where the bale ends meet. Drive your hand into these voids to determine if they go all the way through the wall. If not filled, these gaps create air pockets that weaken the plaster and also provide an avenue for heat loss in cold weather, and heat gain in warmer weather. Filling these gaps creates solid, air-tight walls that will make a home cooler in the summer and warmer in the winter. Packing voids and seams also ensures a solid surface for the plaster.

As soon as the clay slip is applied, and while it is still wet, you must check each and every seam and joint between bales for gaps and crevices. Poke your fingers along the seams and joints between bales to locate holes and gaps in the wall. (Figure 6-7) To fill those that go all the way through the wall, start with dry straw. Use it to fill the "core" or center of the gap. The

Fill holes that go through to the other side with wet mix and dry straw

Pack along posts, windows, and door framing

Wet mix Filled cracks and crevices

Fill cracks & seams with straw coated with a clay slip mix

6-6: Gaps and holes in a straw bale wall are stuffed with straw and straw/clay material. Holes that run entirely through the walls can be first packed with dry straw in the centre, then with 4 – 6 inches (10 – 15 cm) of wet straw clay on each side. Holes do not have to be filled even with the surface, just enough to pack the hole and create a secure base for plaster. The infill coat will be used to fill in remaining depressions.

CEDAR ROSE GUELBERTH

6-7: The areas between bales and the seams where bales meet can create holes that go partially or fully through the wall, creating thermal breaks in your wall system. These need to be filled to build a strong, stable wall system with the best thermal performance.

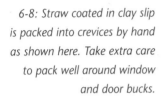

6-8: Straw coated in clay slip is packed into crevices by hand as shown here. Take extra care to pack well around window and door bucks.

inner and outer four to six inches of the gaps are then packed with straw dipped in clay slip (Figure 6-6) This creates an air-tight, solid surface for plaster.

To insert the dry straw, take a handful and drive it to the center of the wall. Use your fingers to push the straw in place. Because you'll perform the same task on the inside surface of the wall, you'll only need to stuff the dry straw in about half way — that is, roughly to the center of the bale. Gloves will help to protect hands.

After the dry straw has been stuffed into the center of the walls, fill the remaining four to six inches with a straw-clay mix (described shortly). Work the straw-clay securely into the wall with your fingers. (Figure 6-8) Don't try to achieve a level surface. Any divots in the wall will be filled later with the infill coat.

In most instances, you'll be stuffing one side of a straw bale wall at a time. Be careful not to push too hard, or you'll shove the straw out the other side. If you let one side of the wall dry before you begin stuffing the other side, there's less chance of pushing straw filler all the way through. In some cases, you will be spraying and stuffing both surfaces on the same day.

6-9: After a section of a wall has been packed, one person should go back over the wall, checking each seam and joint to be sure it has been sufficiently packed, as shown here. Masks are being worn by these workers to prevent inhalation of particles from the slip spray being applied on the adjacent wall.

As noted above, this system is used for continuous holes located at the joints between bales. Along seams or around partial bales you will find gaps that extend only part way into the wall. These can be packed with wet straw-clay only: there's no need to stuff them with dry straw. Be certain to pack well around windows, doors, and posts. Shallower gaps can be filled with infill plaster.

Packing walls with dry straw followed by wet straw-clay results in a very effective, stable wall system. When gaps are packed only with loose dry straw, the straw has a tendency to fall out during plastering. It also tends to create pockets of air behind the plaster, which eventually cause the plaster to disintegrate and fail.

When packing gaps and crevices that extend all the way through a wall, why not fill the entire gap with wet straw-clay? We make a sandwich with dry straw in the center and wet straw-clay on the outside of the gaps because it helps to achieve adequate drying. If the entire crevice were filled with wet straw, the straw-clay filler in the interior would take longer to dry, which could result in decay if moisture remains inside the wall.

As a final note, be sure to assign someone to the role of "master sweeper." This person checks each seam and joint by hand after they've been packed to be sure that none have been missed. (Figure 6-9)

MAKING STRAW-CLAY FILLER. To make the straw-clay filler, dip a handful of dry straw into a bucket of clay slip, or fill a trash can with clay slip and loose straw. You can also spread straw on a tarp and pour clay slip over it, then mix. The longer the straw sits in the slip, the more pliable it gets. For larger packing jobs, it is better to fill a trash can and transport straw-clay to the wall in five-gallon buckets or wheelbarrows.

Dirt screened through a ½-inch screen is sufficient to make slip for this purpose. Some aggregate (sand and pebbles) in straw-clay won't cause any problems. Be certain that the straw is well coated and the clay slip is sticky. If the slip is too watery, it won't adhere as well.

After a wall is packed with straw and straw-clay and the clay slip is dry, you are ready to begin plastering. Although the clay slip needs to be completely dry before plastering, the straw-clay filler doesn't, as long as it has set up.

EQUIPMENT AND MATERIALS FOR MAKING FLOUR PASTE

- Unbleached white wheat flour (preferably organic, high gluten)
- Six-quart pot
- At least one five-gallon bucket
- Water
- Stove/burner
- Whisk or paddle drill
- Optional screen (1/8-inch hardware cloth)

Flour Paste

Flour paste is a binding agent you will very likely use many times when earth plastering. A principal component of the adhesion coat, it is also often added to clay slip and earthen plasters — the discovery, infill, and finish coats — and is used in decorative clay finishes and alises, the colored clay coats painted over the finish coat for protection and visual appeal.

The technique is simple: To make flour paste, first prepare a mixture of cold water and flour, then carefully add it to boiling water while stirring constantly. When it turns translucent, cool the mix and, if necessary, screen it to remove lumps.

Begin with a six-quart pot half full of water. Place it on a stove, covered, over a very hot burner that can maintain a rapid boil. While the water is heating, fill a five-gallon bucket half full of cold water, and slowly add unbleached white wheat flour, preferably an organic, high-gluten flour. (Do not use whole wheat flour.) A high-gluten flour is ideal, because gluten serves as a binding agent. (Note: you can add gluten flour to regular flour to increase the gluten content.)

How much flour do you add? Enough to create a mixture that's similar to a thick pancake batter. Stir with a whisk until your mix achieves the consistency of pancake batter and most of the lumps have disappeared. (You can also stir with a paddle mixer attached to an electric drill.) (Figure 6-10)

Once you've made your flour-water mix and the water on the stove has achieved a vigorous rolling boil, slowly pour the batter into the boiling water. Keep stirring and maintaining the boil. (If the flour-water mix is too thin or is added too rapidly, you may lose the rolling boil; hold off for a few moments to let the flour paste begin to boil again.) If the water maintains its boil, the mix should turn translucent while you are adding the flour-water batter — that is, it should turn from a creamy white milk-like suspension to a slightly translucent suspension. Add enough flour-water mix to the boiling water to achieve a thick, gravy-like consistency. As soon as it has thickened you can remove the mix from the heat.

If your heat source is unable to maintain a rolling boil while you are pouring in the batter, don't despair. This mix will turn translucent if it is boiled longer. (Be careful to continue to stir the mix so it doesn't burn.) You can judge its translucence by taking a spoonful, then letting it stream back into the pot. If it appears slightly translucent, then you're done. Don't expect it to become clear.

Remove the pot from the heat. If your flour paste is lumpy, run it through a window screen into another container to remove the lumps. Warning: flour paste is extremely hot right after it has been taken off the stove and will retain heat for quite a while. Be careful not to burn yourself. Although you can use it immediately, don't scoop it out by hand until it has cooled.

You've now got flour paste, which can be used to make the adhesion coat, added to your clay slip, or mixed with plasters. Flour paste can be covered and stored for later use: it keeps for a day or two without refrigeration (depending on the weather) or a week or two in a refrigerator.

6-10: Mixing dry flour and cool water is made easy with a paddle mixer.

MAKING FLOUR PASTE ON A JOB SITE. Making your own flour paste on a job site can be quite challenging. Most job sites aren't set up to boil water! Kitchen appliances are not usually installed until after plastering. Transporting from a kitchen elsewhere is an option, but for whole house plaster projects, on-site flour paste production is the most efficient and convenient option.

The hotter the burner, the easier it will be to make flour paste. Use a 220-volt electric burner or a high-BTU direct gas burner (sold as heavy duty camp or outdoor cookstoves), either of which can maintain the heat needed to sustain a rolling boil. (Figure 6-11) Small camp cook stoves will not work. When using a large portable gas cooker, be sure to work in a well-ventilated area. It brings water to a boil very quickly, and maintains a boil when the flour is added, which makes production very efficient. (It also keeps workers from making a mess of a brand new household stove — that is, if there's one available!)

Hotplates (110-volt electric burners) can be used, but they generally don't have sufficient power to maintain a rolling boil as the flour-water mix is added. If you use a hot plate, you'll need to cook the flour paste for a while to achieve translucence, which takes longer and increases the likelihood that you'll burn your mix. Be sure to designate one person on your crew to monitor flour paste supplies so that you don't run out of flour paste at a critical time during plaster production.

CEDAR ROSE GUELBERTH

6-11: High-BTU burners can provide the heat to maintain a good rolling boil. If these types of burners are used, batter will cook almost immediately and will be lump-free.

Plastering

Now that you've applied the adhesion coat and slip coats and stuffed the cracks in the straw bale walls with straw and straw-clay, it is time to begin plastering — so long as the slip coat and adhesion coat are dry and the clay-straw packing has set up.

Before we begin, remember that each layer of plaster is designed and applied so that it will key into (attach and bond to) the previous layer. Keying is achieved by mechanical adhesion to roughened surfaces and by chemical bonding of like materials — for example, clay to clay.

Making plaster will require experimentation. We'll provide some guidelines for plaster mixes, but you'll need to make a few different samples of each plaster to assess their feel. After you've made your test mix, take a handful and squish it between your fingers. How does it feel? Does it feel pretty sticky or loose and crumbly? Is it gritty or sandy? Close your eyes and feel the plaster mix. Record your observations.

You'll also need to assess how well it goes on a wall and how well it adheres by applying test patches on small sections of the wall (test sections) or on a test bale. Smear the mix with your hands or a trowel and record your observations.

Applying plaster is easier to demonstrate than to explain in words. So sign up for a workshop if you can! Start with a small handful of plaster, not too much. Hold it in a cupped hand. Then, holding your cupped hand against the wall, push upward smearing the plaster on the wall. (Figure 6-12) In one smooth motion, you should be able to spread the plaster on the wall, using the fleshy part of your hand (between your little finger and your wrist) to flatten it as you apply it. Note that although you start with a cupped hand, you open your hand as the plaster goes on.

Put on a few handfuls, covering a one-foot square section of your test bale or your straw bale wall test section. How does the plaster go on? Does it go on smoothly? Does it stick to your hands? Record your observations in your notebook and mark your test patch clearly: it is amazing how quickly you can forget how you made a particular test patch.

After the test patches have dried, examine them for cracking, hardness, and other properties. If a plaster cracks excessively, it is a sign that it requires more sand and/or straw. If it dusts off the wall or crumbles, it needs more clay and/or an additive such as flour paste or manure. With this in mind, let's turn our attention to the discovery coat.

6-12: To apply plaster, start with a small handful of plaster mix held against the wall, as shown here. Apply pressure and gradually flatten your hand as you move it up the wall. Feather the plaster across the wall surface with the fleshy part of your hand. Keeping fingers together will prevent mud from spilling on the ground.

DAN CHIRAS

The Scratch or Discovery Coat

Slip coat application secures and binds a layer of clay to the straw. Next comes a wet, sticky plaster mix (high in clay) called the *scratch coat* or *discovery coat*. The discovery coat keys into the clay slip coating on the straw bales.

Because likes adhere best to like, the discovery coat has a higher concentration of clay, allowing it to bind securely to the slip-coated straw. Like other plasters, the discovery coat also contains some sand and straw. Flour paste may be included to increase its adhesive properties, and manure may be mixed in to provide fiber and to increase the stickiness of a mix, though both manure and flour paste are optional. The decision to add them depends on the quality of your dirt — its clay content and the characteristics of the clay in the dirt.

The discovery coat is made to the consistency of a thick brownie batter and

MATERIALS AND TOOLS FOR MAKING AND APPLYING A DISCOVERY COAT*

- Dirt sifted through a 1/2-inch screen
- Straw — a mixture of lengths from 2 to 8 inches (5 – 20 cm)
- Water
- Shovels
- Additives (optional) — cooked flour paste and/or manure, cactus juice

- Mixer or wheelbarrow or mortar boat and hoe, for mixing
- Wheelbarrow for transporting mud to wall
- Mortar boards and/or several five-gallon buckets
- Hose and nozzle
- Water source

*Generally applied by hand

is applied in a layer about ¼ to ½ inch thick. It is applied after a very light misting of the dry clay slip coat. Be careful not to mist too heavily or the clay slip will wash off the straw bales.

The discovery coat keys into the clay-coated straw surface and provides a solid base for the next coat of plaster to key into. It also helps to reveal the topography of the wall surface. Remember: the discovery coat is a relatively thin layer of plaster. (Figure 6-13) The next coat, the infill coat, will be thicker and will be applied to fill in low spots and create the final surface you want.

MAKING A DISCOVERY COAT. Plaster recipes vary depending on the soil, climate, construction techniques, and other factors. They also vary from one mudder to the next, and from one site to the next. So what do you do to get started?

If you've read the previous chapters, you should have a solid understanding of the various components of an earthen plaster — what they are and what they do. You should also be able to assess dirt for clay content. With this information, you're ready to start experimenting with plaster mixes.

We can't give you an exact recipe for your location; the mix you'll be using

CEDAR ROSE GUELBERTH

6-13: The discovery coat is applied over the slip-coated bales. This coat applies quickly and easily, creating a solid base for the next coat of plaster. This coat reveals the topography of your wall, hence the name "discovery coat."

depends on the subsoil you have and its characteristics — that is, its ratio of clay, silt, and sand; and the type of clay it contains. However, we can provide a starting point for experimentation. To begin, try one part clay-rich soil, one part sand, and one half-part straw (up to six-inch lengths). In some areas, all you will need to do is mix subsoil and appropriate amounts of straw because there's enough sand in the soil to make a good mix.

Use a shovel or a bucket to determine the parts, though using buckets usually produces a more consistent quality. If you're using a shovel, be sure the same person does the shoveling. Before you begin mixing your components, screen your materials, as explained in Chapter 5.

When mixing, begin with a small amount of water, then add half of the sand. After the sand and water are mixed, add half of the sifted dirt and mix. Once this is well mixed, add the other half of the sand, mix, then add the remaining dirt, mixing thoroughly as you go. When these ingredients are well mixed, add the straw or other fiber.

While mixing the plaster for your discovery coat, you may find that the mix gets too dry and doesn't mix well. If this occurs, add more water — but go easy. You'll find that the line between being too dry and being water-saturated is quickly overstepped. Many novice mudders are amazed to find, after adding a little water to their mix, that a plaster which appeared too dry suddenly becomes wet and sloppy. If this happens, you'll need to add more sand, clay dirt, and straw in the same ratio as the original mix — in this case, to create the consistency of a thick brownie batter. Be aware that straw will absorb some moisture. Mix thoroughly.

Once your plaster is mixed, it is time to apply some test patches, a process described earlier. Remember: waiting for each test batch to dry can take a while, so you may want to create several test batches initially. By changing the ratio of clay, sand, and straw in your initial mix, you can create several samples, which can be applied in the same afternoon and assessed a day or two later. In one batch, you may want to add more dirt. In another, you may want to add more sand. In still another you may want to add more straw. In another test batch or two, you may want to add cooked flour paste. Yet another might be made with sifted manure or a combination of sifted manure and cooked flour paste or other additives.

Apply each test mix to the wall or to a test bale sprayed with slip. Be sure to mist the wall or test patch before applying the plaster. Spray very lightly with water to dampen the surface: don't spray too heavily or the slip will run off. After applying each test batch, be sure to note how it goes on the wall. Is it stiff enough to cup in your hands but loose enough to spread easily and adhere well? Does it go on smoothly? Does it adhere well? Does it key into the bales well? Now let them dry so you can compare the finished product. Does the plaster adhere well to the wall? Does it crack? How extensive is the cracking? Hopefully, you should be able to come up with a mix without too many tests. As you gain experience, this process will require fewer tests.

If the plaster does not adhere well to the bales, the mix may be too dry, may not contain enough clay, or may require natural additives. Start experimenting with a cup of flour paste and/or a half a shovelful of sifted manure per five gallons. For the discovery coat, your primary goal is a plaster that adheres and keys into the bales. If the test batch crumbles off, that's not a good sign: it means your plaster doesn't contain enough clay.

> ## APPLYING THE DISCOVERY COAT WITHOUT CLAY SLIP
>
> Many straw bale builders apply the discovery (scratch) coat directly on straw bales without a clay slip coat, but find that the plaster tends to peel off during application. This approach results in a weaker bond between plaster and the straw bale. It's also a lot harder work! For these reasons, we highly recommend applying a clay slip first.
>
> If circumstances require application without clay slip, be sure to mist the bales first. The plaster mix should be very sticky and spreadable so it will key well into the surface. This may take more time and effort, but a stable base can be provided for the next coat.

Don't worry too much about cracks in the discovery coat; some cracking is fine so long as the plaster adheres well to the wall. In fact, cracks actually help the infill coat (the next layer) key in more securely.

APPLYING DISCOVERY COAT PLASTER. After you have found a mix that works for you, it is time to begin work. Don't forget to organize the job site so you can work efficiently with a minimum travel distance between materials, mixing, and application. Make sure the site is safe, too. (Both topics were discussed in the last chapter.) Prepare as many of your materials as possible in advance, or have a well-organized crew to stay on top of material preparation.

Now it is time to begin plastering your walls. Begin by lightly misting the wall with water. Use a hose with a nozzle that permits a fine mist. Anything heavier than a fine mist, and you'll wash off the clay slip. Another option is a hand-pump garden sprayer. Buy a new one; don't use one that has had pesticides or fertilizer in it. The mist will very likely evaporate quickly, especially in hot, dry climates, so spray a small section at a time.

The discovery coat is usually applied by hand. You may want to wear rubber gloves to protect your hands if you find the straw and sand in the plaster are too abrasive. Keep a bucket of water nearby to clean your hands from time to time; you'll be glad you did.

As in the test patches, apply the discovery (scratch) coat one handful at a time. To begin, take a small handful and hold your cupped hand against the wall; then push upward, opening your hand as you go, and smoothing the plaster with the fleshy part of the hand between the base of the little finger and the wrist. This part of the hand seems to be custom-made for earthen plaster application! If plaster is falling on the ground around you, you're most likely not cupping your hands and/or trying to apply too much at once.

While plastering, focus on keying the mix well into the surface of the bales (see Figure 6-12). You'll need to apply pressure as the plaster goes on the wall, but not too much. A little experimentation will give you a sense of the right amount of pressure. Too little and it won't key in well; too much and it goes on irregularly and spills onto the ground. You may need to go over some areas a second time to achieve a more uniform layer, but as you become experienced, this probably won't be necessary. The discovery coat should go on relatively quickly.

The discovery coat is applied in a layer about ¼- to ½-inch thick. Remember, it is not used to even out or shape the surface. It is simply applied evenly over the entire surface and is keyed into the clay slip coat, revealing the topography of the wall. (Figure 6-14) Some mudders may scratch the base coat with a cement stucco tool, called a *scratcher*. This small rake-like device is run across the wall parallel to the ground. It cre-

6-14: This photo shows the top portion of a straw bale wall coated with the discovery coat. The discovery coat should be only about one-quarter to one-half-inch (0.65 – 1.3 cm) thick, applied by hand and keyed into and evenly over the surface of the clay slip coat on the straw bales, revealing the topography of the wall. The lower section of the wall shows the next (infill) coat, which evens out to create the surface you want.

ates small grooves that will make the next coat adhere more easily. As noted earlier, this is usually not necessary because hand application of the discovery coat produces a rough surface, which makes a great key for the next coat, the infill coat. (For large projects or commercial applications, plaster can be applied by machine, such as a pneumatic pump-sprayer. We'll discuss this option at the end of the chapter.)

Infill Coat

The discovery coat must dry thoroughly before the infill coat can be applied.

The infill (or brown coat) allows you to define the final shape or, more precisely, the desired surface topography of your walls. The infill coat contains the same components as the discovery coat, except that it contains more sand and straw (two- to eight-inch strands) to provide structural strength. It will be thicker, too, about one-half to two inches in most cases, although it could go on up to six inches. We'll talk some more about that in a few moments.

MAKING THE INFILL COAT. Once you have found a successful mix for your discovery (scratch) coat, determining the proper ratios of clay-dirt, sand, and straw for your infill coat is much easier. If you used clay-dirt in a ratio of 10 parts clay-dirt to 10 parts sand and 1 part straw in the first coat, the second coat may be 10 parts clay-dirt to 15 parts sand and 3 parts straw. Screen materials before making a test batch: ½-inch screen for dirt and ⅛-inch screen for sand, unless it is prescreened from a gravel pit.

The infill plaster should be slightly drier than the discovery coat mix. You'll mix it the same way, however. Begin by adding water to the cement mixer (or wheelbarrow or pit), then add half of the sand. Mix, then add half of the clay-dirt and mix some more. Then add the remaining sand. Mix, then add the rest of the dirt. Add manure or flour paste (if needed), and finally add straw and mix thoroughly. As in the discovery coat, add water if the mix appears too dry, but remember this mix should be slightly drier than the discovery coat, while remaining pliable and workable. Once again, be careful not to add too much water! If your mix is too wet, you will need to add more ingredients in the proper ratio.

Now comes time to test the batch. Take a handful of the material, and give it a good feel. Although it still contains a considerable amount of clay, it should feel grittier than the discovery coat plaster because it contains more sand, which provides structure and bulk.

Be sure the discovery coat is thoroughly dry before applying the infill coat. Wet the surface with a good spray nozzle, but be careful not to apply so much water that it runs down the walls. Once the wall has been wetted, wait a minute or two so the water can soak into the surface, then apply the infill plaster test patch over the discovery coat.

The infill plaster should adhere well and should allow you to build out and to shape a wall. If not, you may need more sand. Try a second batch and a third batch, each with more sand, if necessary, but let them dry before proceeding. Be sure to keep track of (record!) the changes you made and your observations.

MATERIALS AND TOOLS FOR MAKING AND APPLYING THE INFILL COAT*

- 1/2-inch screened dirt
- Sand
- Straw (combination of lengths, two to eight inches)
- Water
- Shovels
- Cement mixer (or mixing boat or wheelbarrow with hoe)
- Wheelbarrow for transport of mud to walls
- Mortar board
- Five-gallon buckets
- Trowels

*A combination of hand and trowel application is typical

122 THE NATURAL PLASTER BOOK

After the wall has been wetted, add a little water to a handful of infill plaster mix. Spread a thin film across the surface to "engage" the discovery coat. Then begin applying the infill coat to build out the wall.

After the plaster test patches have dried, assess the level of cracking. Remember, excessive cracking is not good at this stage. If an infill plaster sample cracks excessively, you'll need to add more sand. If the plaster is crumbly or doesn't adhere well, it probably contains too much sand — in which case, add more clay dirt or add flour paste and manure. Apply another test batch and see how it goes on and how it looks after it dries.

APPLYING THE INFILL COAT. Once you've determined the composition of your infill plaster, it is time to get to work. You'll need to mix large quantities of plaster.

Before you begin to apply the plaster, be sure to wet the discovery coat. Once again, don't waste time spraying an entire wall at once if you are working alone or with one other person: it will usually dry before you get to it and will require re-spraying. If you're working with a larger group, you can wet much larger wall sections, but be sure your plaster mixing crew is working full speed to keep up with applicators.

The infill coat is typically applied by hand. In fact, hand application is preferable. Run an initial thin coat of infill plaster on a section of wetted wall, being sure to key it into the discovery coat. This coat should be about ¼- to ½-inch thick. After it has keyed into the discovery coat, come back and add more plaster to build out the wall — that is, to shape or smooth the surface.

As noted earlier, you may need to add two to six inches of infill plaster, the latter if you encounter deep holes. In such instances, you may need to apply the plaster in a couple of layers, probably no more than two to four inches at a time, feathering it as you go. (A drier mix allows for more build out — that is, thicker layers.) If your mix is wet or you have added too much plaster at one time, the plaster will slump and could peel off the wall. You can tell when earthen plaster is getting too thick by its appearance: it starts to droop while still wet. If this occurs, stop and let the plaster set up. (As a general rule, drooping begins after you've applied up to four inches of plaster.) After the plaster dries and sets up a bit, apply another coat. Continue applying infill until the wall attains the topography you want. Infill plastering consumes a lot of mud plaster, so be sure you have several people prepping materials and cranking out plaster while the mudders are applying the infill coat.

Hand application of infill plaster can be time-consuming, depending on the condition of your walls. The infill coat can also be applied with a stucco machine (discussed at the end of the chapter). Although machine application speeds up the process and results in a fairly tight bond, there is still a considerable amount of hand work be done — for instance, you'll need to have a several people following the applicator to spread the mud out either by hand or with trowels.

More skilled mudders should apply the plaster around windows, doors, corners, and around electrical boxes for outlets and switches. If you want straight, square corners, you will need to use two trowels, one on each wall. (Figure 6-15) Mudding around window and door frames may require the use of screws for the plaster to key into (see Figure 3-20). The details of applying plaster around doors and windows are described later in this chapter.

As you shape a wall, you may want to use a trowel. But don't trowel the surface too smooth: leave the surface a little rough to allow the finish coat to key nicely into the infill coat. A light brushing with a broom — or going over the surface with your hand — will roughen the surface enough to enhance adhesion of the finish coat.

If you're artistically inclined, you can use the infill plaster to begin sculpting features — faces or animals or flowers. (Figure 6-16) Remember when working outside, work in the shade to prevent plaster from drying too rapidly and cracking. Some cracking is to be expected and occasional cracks in this layer are okay — so long as they are not excessive and are not causing plaster to peel away from the wall. Cracks allow the next plaster coat to key into the surface. As long as they are stable and secure, cracks can be filled as you apply the next (and final) coat.

THE IMPORTANCE OF FEATHERING THE INFILL PLASTER

As you apply infill plaster to a straw bale wall to fill the holes and depressions, remember to feather the plaster. Begin by applying a handful of plaster to a depression. As you work the surface to key in the plaster, feather the edges outward (taper the thickness of the plaster), pushing with sufficient pressure to secure the plaster to the wall. You can now apply another layer on top of the first one, if needed to fill the depression. Keep working the mud until you feel confident that it is well keyed into the wall. This technique allows a mudder to produce a more even wall surface and to soften curves, as well as providing a secure foundation for the next layer of plaster.

If you add too much infill coat at any one time or if the mix is too wet, the plaster is likely to slump (droop or sag) and may even peel off and splatter at your feet. When you need a thick layer (three to six inches), apply it a couple of inches at a time, and allow each one to set up well before applying the next.

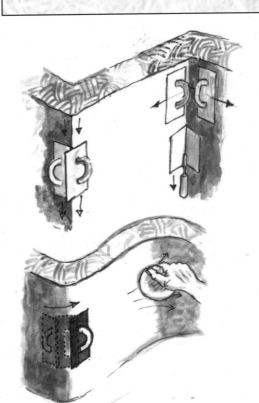

6-15: To form sharp, square corners use two trowels held in a V as shown here. This will take a little practice. Or if you want, corners can be rounded. Inside corners can be achieved by using a bent plastic disk or a plastic or rubber flexible corner tool.

The Finish Coat

After the infill coat has dried, it is time to apply the finish coat. The finish coat serves several important functions; most notably it provides beauty and durability. The beauty of a wall stems from its topography (shape), color, and texture. The durability results from its ability to seal a plaster wall, which is why the finish coat should have a higher proportion of clay and less sand than the infill coat. Additives such as flour paste, cactus juice, and manure also help create a more durable and more tightly sealed surface.

Don't forget, the topography of a wall is largely determined by the infill coat, although the finish coat can also smooth out some minor bumps and dips. Color is a function of either the natural color of the clay you're using in the finish plaster or pigments added to it to provide color. (You can select dirt for your finish coat based on color, so long as the dirt contains enough clay.)

DAN CHIRAS

Texture is a function of the way the finish coat is applied and finished after it is on the wall. You'll find that there are many different textures to choose from, ranging from a rough porous look to a smoothly polished surface, and a number of different tools and techniques to achieve the desired effect.

The finish coat should not dust off a wall when a person or a pet brushes against it; rather, it should have a hard, durable surface. Natural additives help increase the hardness and durability of the finish coat, and although there are many to choose from, we have found that cooked flour paste is conven-

6-16: Earthen plasters are ideally suited to wall sculpting, producing delightful figures like this dolphin or more complex sculptures like the one on the cover of this book.

ient, as it is made from readily available materials and is relatively simple to mix. (Casein, powdered milk, and cactus juice also work well!)

MAKING A FINISH COAT. The finish coat is made from the same ingredients found in the discovery and infill coats: clay-dirt, sand, flour paste, manure, and sometimes straw. You can also make it from powdered clay, as we explain shortly.

When working with clay-dirt and sand, remember that all materials should be sifted: the finer the screen, the finer the finish coat. Mix the same way as other components. Start with water and half of the sand; then add half of the clay-dirt. When mixed well, add the rest of the sand; mix thoroughly and add the rest of the dirt. Flour paste and manure can be added anytime. Straw — which is optional — is added last. If you do use straw, make sure that it is finely chopped — into strands from one to two inches in length.

Remember that in some climates, straw in an exterior finish coat may cause the plaster to erode away. Use manure instead, as it contains very fine fibers that give the plaster tensile strength. It won't create problems with erosion, and contains enzymes that make the finish more durable. Add water as needed.

As before, you will need to make some test batches and apply them to a plastered test bale or a plastered test section of the wall to discover the optimum mix. The mix for a finish coat should contain enough clay for adhesion and protection, and enough sand and fine fiber (manure or wool) to create a stable surface with little or no cracking. Adding flour paste, cactus juice, powdered milk and other natural additives provides extra durability and prevents dusting. The mix should be slightly stiff, pliable, and easy to apply.

APPLYING A FINISH COAT. Before the finish coat can be applied, be sure the infill coat is completely dry. The wall surface must be wetted before a finish coat is applied. Use a spray nozzle or daub or splash water on the wall with a large paint brush (6 inch or 12 cm) or a sponge. (Brush and sponge application can be really slow.) Remember, don't apply so much water that it runs down the wall — and wait a few minutes for the water to penetrate the surface before applying the finish coat.

Finish coats can be applied by hand or by trowel. Trowel is the most common and easiest method of application. You'll be applying a layer about $1/8$-inch to $1/4$-inch (0.3 – 0.6 cm) thick. If you're applying the plaster by hand, hold a small handful in one cupped hand, then smear it onto the wall, working from the bottom of the wall up. Try to apply the plaster uniformly and evenly. Hand application produces a rough finish coat and results in a more porous surface that is more vulnerable to weather and thus more likely to deteriorate than a hard-troweled finish.

Troweling a finish coat generally yields the smoothest and most long-lasting finish. If you are using a trowel, place a small amount of plaster on the blade of the trowel, then apply it to the wall by angling the trowel so that the plaster goes on smoothly and the leading edge and sides of the trowel don't slice into the wall. (Figure 6-17) You'll need to apply some trailing edge pressure, but not too much. Work the plaster with even, consistent strokes in different directions to achieve a durable finish.

MATERIALS AND TOOLS FOR MAKING AND APPLYING A FINISH COAT

- 1/2-inch screened dirt
- Fine sand
- Fiber (cut hemp or wool or sifted manure)
- Optional additives: flour paste and/or manure, cactus juice, casein, powdered milk
- Optional pigments
- 1/4-inch screen
- Trowels
- Bucket of water
- Five-gallon buckets
- Mortar boards
- Hawk (optional)
- Sponges, burlap, and/or plastics disks (for polishing walls)

When working with a trowel, be patient. Don't try to cover large sections of the wall with each trowel load of plaster. Take your time: put small amounts on the trowel and spread evenly and well. As you get better you can apply larger amounts, but while you're learning, you'll just end up spilling a lot of plaster on the ground. You will very likely need to go over each section a couple times to smooth out the plaster very soon after you first apply it. Again, be patient — but don't wait too long to smooth out the finish coat as it will set up and become more difficult to work. Be careful not to overwork the finish coat,

 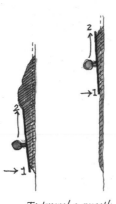 

Apply plaster to the wall by hand or trowel, or a combination of both

To trowel a smooth, long-lasting finish
1) Apply pressure to the trailing edge of the trowel, while you...
2) Move the trowel forward

Your trowel pattern may vary depending on the look you want your wall to have.

6-17: Plaster application by hand or by trowel.

too: it may crack if you do. Keep a bucket of water handy to clean the trowel from time to time; it makes the work easier and more pleasant.

After the finish coat has dried to the hardness of leather, you may find small dips or depressions or trowel marks in the wall surface. Buffing the surface with a slightly moist sponge or slightly moist piece of burlap can help minimize or eliminate them. However, too much moisture can produce water stains and too much rubbing can mar the surface (Figure 6-18). Be sure to buff the entire wall to produce a consistent finish.

6-18: Walls can be finished by buffing wall with (a) a slightly moist sponge or burlap, or (b) a hard steel trowel, or (c) a plastic lid to even out the surface and produce a hard, durable finish. See text for details.

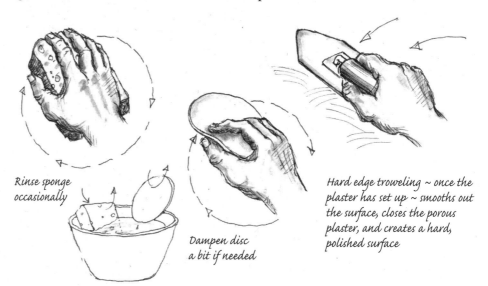

Rinse sponge occasionally

Dampen disc a bit if needed

Hard edge troweling ~ once the plaster has set up ~ smooths out the surface, closes the porous plaster, and creates a hard, polished surface

TEXTURIZING A FINISH COAT. The surface of a finish coat can be texturized in one of several ways. These techniques allow one to produce an aesthetically appealing wall surface that provides a variety of functions as well. Three basic surface textures are possible. First is the open finish, which is highly porous — it looks rough and is rough to the touch. It can be achieved by trowel or hand application.

The second is the closed finish, and is achieved by troweling or buffing the wall after the surface has set up — that is, after it has achieved a "leather" dryness. Hard steel trowels — for example, high-quality flexible Japanese stainless steel trowels or small steel pool trowels — work well. Apply pressure to the trailing edge of the trowel as it is worked in several different directions. Beware: overworking a wall or working it when it is too wet can cause cracking.

A closed finish coat can also be achieved by buffing with a plastic lid (a yogurt lid works well) or a slightly moist (well-wrung-out) sponge. When buffing with a plastic lid, carefully rub the wall using circular motions to smooth out the surface of the hardened plaster. You may need to dip the lid in fresh water from time to time to make the buffing go a bit more easily. Be careful, though: too much water will cause the finish coat to crack or stain. Such treatments seal the surface pores and create a continuous smooth, well-sealed surface.

The third option is a "cat face" finish, a combination of the open and closed finishes which requires a fair amount of practice to achieve. After the finish coat is applied, before the finish sets up, go back over it, troweling across the surface, leaving drag marks and open patches. Run the trowel across the surface of the plaster, applying some pressure. As you do, you'll notice that some areas of finish coat are rough and some are smooth, and that drag marks and patches appear on the surface: these are known as "cat faces."

ADDING COLOR TO YOUR WALLS. The natural color of the dirt in the finish coat is often quite appealing to many people. Another option for providing color is to add pigments to the finish coat or to apply a finish coat made with white clay plaster (for example, kaolin clay) that is tinted with natural pigments (discussed in Chapter 7). A colored lime plaster can also be applied instead of an earthen plaster finish coat, as explained in Chapter 8, as can a colored gypsum plaster (Chapter 9).

Finish coats can also be painted — that is, coated with one of several natural "paints." For example, earthen finish coats can be painted with alises, which will be discussed in Chapter 7, along with several other options, including silica paints, lime washes, and milk-based paints.

Clay-Straw Plaster

Bill and Athena Steen, authors of the best-selling *The Straw Bale House*, have been instrumental in making straw bale building popular, and have devised many techniques that help make it better, too. One technique is their clay-straw plaster, a two-coat system that works well in arid climates.

As the name implies, clay-straw plaster is primarily a mix of clay slip and chopped straw. Working with people in Mexico, Bill and Athena devised a plaster recipe that seems to work well on straw bale walls in hot, dry climates. To make the plaster, they mix one part of clay slip made from local soil, about the consistency of a milkshake, with four or five parts chopped straw, and, occasionally, add one or two parts sand. The ideal mix should be sticky and should hold together pretty well. You don't want a lot of clay slip oozing out of the mix. Try making a few samples using different ratios of straw to obtain the perfect mix for your job.

Clay and straw plaster is sticky and malleable, and has high tensile strength. It mixes well either by hand or in a mortar boat, and can also be mixed by foot in a tarp-lined pit. Start with some straw, add clay slip, then stomp. From time to time, you will need to mix the material by hand to ensure that the clay slip is coating all of the straw.

Clay and straw plaster makes for a great fill coat and can be applied quite thickly over a straw bale wall. The mix is often tossed onto the wall and worked into the surface, then built out to substantial thickness (Figure 6-19). The straw creates a rough surface, which allows the next layer (the finish coat of clay plaster or lime) to be keyed in quite effortlessly.

6-19: Bill and Athena Steen demonstrate their clay and straw plaster techniques applied in desert climate.

BILL AND ATHENA STEEN

Interior vs. Exterior Plasters

Interior and exterior plasters live very different lives, one potentially harsh with wide temperature swings; the other, a more refined and gentle existence with relatively constant temperatures. Interior plaster doesn't get rained on or receive a daily beating from the sun. Because of this, the interior and exterior plaster should be made slightly different.

To begin, for protection the outside plaster should be at least two inches (5 cm) thick. To protect the outside plaster, some mudders stir a little linseed oil into the finish coat mix (one tablespoon per 5 gallons). Too much oil, however, tends to break down and can cause the finish coat to peel off. It also prevents easy repair down the road and refinishing.

A slightly higher clay content is advisable for exterior plasters, but not so much that it causes cracking. Manure increases the durability of exterior plasters and prevents cracking. Flour paste or cactus juice (about 1/2 cup per 5 gallons) mixed into the exterior finish coat also creates a more durable finish. Whatever you do, be sure to test your finish coat first.

As a final note on the subject, flour paste, cactus juice, and extra clay create a durable finish that can expand and contract without cracking as the temperature changes. Although these additives can provide increased durability, they are not as necessary for interior plasters, which inhabit a milder environment than exterior plasters. However, flour paste and other additives will prevent dusting of a finish coat and are therefore often added to interior finish plasters. Hard troweling and buffing a wall can also prevent dusting.

6-20: Exterior plasters are exposed to much harsher conditions. A slightly higher clay content and additives such as manure or flour paste mixed into the final coat create a more durable finish that can expand and contract in response to temperature fluctuations without cracking.

Plastering Around Framing, Windows, and Doors

Straw bales and earthen walls often abut framing, such as exposed timber frames and exposed window and door frames, or abut other building materials, such as stone, adobe, and concrete. Special care needs to be taken in such places. When straw bales abut framing members or other materials, it is important to pack the spaces between the wood or other materials and the bales with a straw/clay mix, described earlier in the chapter. Let it dry, then finish by filling in the area with a wet, sticky cob-like mix to shape the surface. (Figure 3-19). Allow it to dry, wet the surface, then apply the discovery coat. Use the infill coat to shape the area. Build out in several coats and fill and pack cracks between the wood frame and packing material with plaster as needed. Be sure each consecutive coat of plaster keys

CEDAR ROSE GUELBERTH

6-21: An infill plaster (a cob-like mix) can be used to build out and sculpt around well-packed window and door frames to create soft smooth lines, as shown here.

tightly into the gap between the two surfaces. By the time you arrive at the finish coat, you should have eliminated any signs of cracking. When packing against large spans of wood frame, you can provide added strength by driving screws into the wood, leaving ¼- to ½-inch heads exposed. This provides a nice surface to key the mud into (see Figure 3-20).

Protecting Plaster from Rain and Freezing

Plaster needs to be protected from rain during application (when it is most vulnerable) and when it's drying. It also needs to be protected from freezing temperatures: if a wet plaster freezes, ice crystals cause it to expand, weakening it. When it thaws, the plaster contracts, causing further damage. Stop working well before ambient temperatures are expected to drop below freezing. When plastering in questionable weather, walls should be covered with tarps at night to reduce the chances of freezing. (Be sure to mount tarps away from walls so as not to trap moisture.) Better yet, don't apply earthen plaster if there is any risk of freezing. Cold weather, even when freezing is not a factor, can also cause problems. For example, it slows the drying of a plaster and may prevent it completely — for example, during long, cold, wet winters.

Tips on Applying Plaster

Plastering is fairly simple, but there are some bits of advice we can share that will make the job go more smoothly. First, be sure to follow the guidelines described in Chapter 5 on safe and efficient plastering operations. Second, be sure to apply the prep coats. Without a well-penetrated slip coat, earthen plaster tends to peel off straw bales. In addition, without a layer of clay slip, you'll need to work harder to key the discovery (scratch) coat into the straw bales. The adhesion coat is also important when trying to key plaster onto uniform flat surfaces such as wood or metal plates. Third, as we've said many times before, be sure to organize your work site for efficiency, in every step of the process from preparing your materials to mixing plaster to transporting it and applying it to your walls. Fourth, be sure to make plastering safe. A trip to the hospital can really cut into your work time, not to mention that it's not an enjoyable experience for the wounded mudder!

PLASTER WITHOUT PAIN

The key to plastering a wall is to use the mechanics of your body to decrease strain on muscles and joints. Plaster is heavy and thick. An expert mudder makes it as easy as possible on his or her body.

Our fifth tip is to tape windows, door frames, and exposed woodwork, stone, and tile and any other surface you want to protect from plaster splatter, which can stain wood and other materials. You may also want to cover the floor with cardboard or canvas tarps, though, ideally, finished floors should not be installed until after plastering is complete.

Sixth, drying occurs quickly in many climates, even in wet climates during the summer months. Although you want discovery and infill plaster coats to dry completely before applying the next coat, you don't want them to dry too quickly. If they do, some cracking may occur. Avoid application in direct, hot sun. In the morning, for instance, you can plaster the west side of a house. During the afternoon, shift operations to the north and east sides. As the sun starts to descend, tackle the south side — or save it for a cloudy day.

In the system we're recommending, the slip coat is applied by machine (usually a drywall texture gun and compressor). The discovery coat is applied by hand. The infill is applied by a combination of hand and trowel. Be sure any troweled surface is gone over by hand or brush to create a rough surface for better adhesion of subsequent coats. A stucco machine also works well for the discovery and infill coats. Feather plasters as you're applying them.

The finish coat can be applied either by hand or trowel; often a combination of the two works fine. That is, you may want to put the material on initially by hand, then smooth it out with a pool or a Japanese steel trowel. Remember, hand application results in an open, porous, rougher look. Troweling, on the other hand, closes pores and seals the surface, leading to a longer-lasting finish. Be sure to work the trowel in different directions.

Another important tip is to watch for waste. Veteran plasterer Bob Campbell reminds his workers and our readers, "You've gone to considerable trouble mixing all this stuff. So let's get it where we want it. If it is all over you and on the ground, it is not going on the wall." Campbell goes on to say, "The tools of plastering are pretty primitive, but they allow incredible flexibility when you know how to use them." Beginners can develop control without a huge investment of time.

Professional plasterers typically use a hawk to hold plaster, which is then transferred to a trowel (Figure 6-22). This is achieved by a hook motion. Begin by pushing a gob of mud up on the hawk, which is tipped back and away from you. Then, scoop some onto your trowel while angling the hawk back, as illustrated. (Hawks are rather cumbersome and difficult to use. Because beginners have to struggle with them and often feel a considerable amount of frustration, the hawk is listed as an optional tool.)

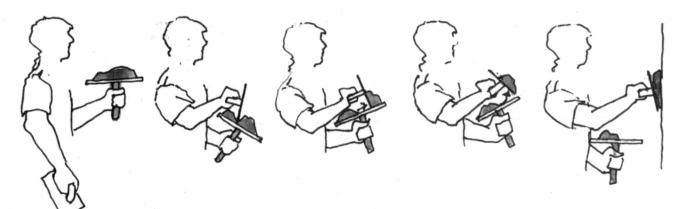

6-22: Plaster is held on a hawk and removed by maneuvering the trowel and the hawk simultaneously. Begin by pushing a gob of mud upwards on the hawk, which is tipped back and away from you. Then scoop some onto your trowel while angling the hawk back toward you. This is achieved by a hook motion. Then scoop some onto your trowel while angling the hawk back.

6-23: To prevent mold forming on freshly plastered walls, open doors and windows to ventilate the room. Circulation fans, shown here, may help. Running heat in colder or humid climates will speed up the drying of plaster, as well. In especially humid climates, a dehumidifier may help eliminate the problem.

Another bit of advice on plastering: when working with large groups, assign — or ask workers to claim — segments of the wall. It is sometimes hard to tell what's been recently plastered, especially if you are working in the shade: assigning wall segments minimizes duplicated effort. Be sure that all mudders are applying plaster in the same manner, and working carefully to create a consistent finish. When applying finish coats, identify your work area for the day and be sure that you finish a complete wall section or stop at a corner to avoid obvious seams where one day's work ends and another begins. Designate a quality control person to review all work each day.

On a final note, plastering is often done during the summer with windows open for good ventilation. In the winter, however, earthen plaster may be applied while a house is closed up, and humidity may cause mold to form on the walls. If this is the case, you'll need to take steps to improve air circulation. You might, for instance, operate a fan or two and open some windows during the warm part of the day, or you could install a dehumidifier to extract moisture from the house. In humid climates, cross ventilation and fans may not be enough to thwart mold growth; adding a dehumidifier will often solve the problem. Mold can be eliminated by spraying the mold with hydrogen peroxide. However, this is only effective when used in conjunction with increased air circulation and ventilation (Figure 6-23).

CEDAR ROSE GUELBERTH

Plaster Parties and Workshops

Plaster parties are a great way to mud a house quickly and economically. If you take this route, be sure that everyone is trained adequately first. Moreover, be sure that someone is present who can manage the crew, supervise the work, and answer the many questions that arise over the course of a day. (You may want to reserve the final infill coat and finish coat application for experienced plasterers.)

One way to get the job done is to hire an experienced local or national earthen plaster expert to consult with you. That person can help you decide on the proper ratios for your plaster mixes and can offer instruction and guidance for your group. As well, you could hire an experienced plasterer to train and/or manage a crew.

Another option is to sponsor a workshop. Organizing workshops takes a lot of time and effort, and the amount of plastering a group of novices can accomplish is usually minimal unless you sponsor an extended workshop lasting ten days to two weeks. In addition, there's the issue of quality: people who are just learning plastering are unlikely to do as good a job as a team of experienced mudders. You may not want to compromise quality for the promise of temporary free labor. Training your own crew or hiring an experienced professional, if you can find one, may be a better option, producing the best results. Fortunately, if you are using your own crew, you may find that your crew is pretty experienced by the time they finish the discovery and infill coats.

Hand vs. Machine Application

As noted in earlier sections, earthen plaster can be applied by hand or by machine — for example, a Quick Spray pneumatic spray applicator. These machines, which consist of a hopper, a pneumatic pump (peristaltic action), hose, and nozzles which blow a wet mix onto walls, cost $5,000 to $7,000, and you'll need to rent or buy a compressor, too (Figure 6-24). Special nozzles can be purchased to apply mud plaster. Some plaster or stucco application equipment comes with mixers that feed the gun, but the price of these units is prohibitive for all but professional crews with lots of work lined up.

Machines represent a sophisticated, quick, but expensive way to move mud from the wheelbarrow to the wall. Don't be lulled into complacency here: using a machine doesn't eliminate the hand work. You'll still have to prepare and mix plasters to feed into the machine, and do it much more quickly. In addition, you'll need a team to trowel the plaster after it is sprayed on the wall to smooth it out. If you're working alone, machine application makes no sense. If you are running a crew, it may be a great option — especially if you plaster a lot of houses each year.

Machine application works well, but for one time owner-builder, it can be a fairly expensive. If you can borrow or rent a unit, or even buy one then sell it when you're done, it might be worth the initial cost. It all depends on the size of house, the size of the crew, the time constraints, and the size of your bank account. If you're in a bind for time — for example, winter may be fast approaching or your construction loan is racking up huge amounts of interest — the investment, either rental or purchase, may justify the expense. Some professional crews use machines to apply plaster, which saves a considerable amount of time. Although you may be able to save some money by being able to complete a project faster that way, hand application can save money with volunteer or cheap labor. Carefully consider the cost of each option.

6-24: Workers applying mud plaster using a pneumatic pump sprayer.

Besides speeding up the process, machine application really keys a plaster into bales well. They're powerful and can do a great job of ensuring adequate keying of the various layers of plaster, assuming a well-trained crew is making the plaster, operating the equipment, and applying it.

Hiring a Professional Crew

In their book, *Straw Bale Building*, Chris Magwood and Peter Mack write, "A natural division seems to exist in the human population: those who enjoy plastering and those who detest it. If you are among the former, you might actually have a good time putting in the effort that's required. If you are among the latter, it may be worth any price to have professionals do the work for you!"

There are a few earth plaster crews that will travel to your site and plaster your home, but not many. Contact local natural builders for recommendations. Be careful who you hire: there are good crews and, well, some not-so-good crews. Ask a lot of questions. How long have they been in business? How many homes have they plastered? Were they the same type of home you're building? What types of climates have they plastered in? How much experience do they have with earthen and other natural plasters? Where did they learn how to plaster? From whom? Will they supply you with references from jobs that went well and jobs that didn't? Remember that references from jobs that go well are great, but it is very important to understand how a crew handles problems and mistakes that inevitably occur during construction. A responsible crew will take the responsibility to come up with solutions. Ask other natural builders if they know of the plaster crew you are considering. When talking to references, find out how the crew performed. Did they show up on time? Did they complete the job well and on time? Were they reasonably priced? (Remember: cheapest isn't always the best.) Were they honest? If a job went wrong or problems were encountered, did they make amends? (That's a good referral as far as we're concerned.) Remember, some jobs go poorly, but if the mudders made it right, then they may be considered reputable and trustworthy. Ask them what they do when mistakes occur, and what kind of problems could arise after they've completed the job. What would be their responsibility and what would be yours?

It is hard to get guarantees for earthen plaster jobs because plastering is a tricky business. A good plaster job on a poorly designed, engineered, and constructed home won't hold up — and nine chances out of ten the mudder will be blamed. Because the quality of a plaster job depends on these details, a mudder should be consulted during the design stage for feedback. Many people don't understand that buildings shift and settle over time, and that expansion and contraction of building materials can cause cracking in the plaster finish. Homeowners should be aware of this before plastering begins. A plaster job that doesn't hold up to settling and other forms of movement is very likely not the fault of the mudders, as a rule. Unless the applicators have done a really bad job of plastering, cracking is usually the consequence of design and construction and naturally-occurring settling. Note that bad plastering results in immediate cracking; cracking caused by settling usually appears some time after the plasters have dried.

Even a well-engineered building will experience minimal cracking from settling. You could choose to wait a year or two before applying the finish coat, or you could apply an alis finish to fill and refresh plaster walls when hairline cracks appear. Larger cracks can be filled with matching plaster or filled and coated with alis. (Note that large cracks or continual cracking may be an indication of structural problems.)

Quick Recap

In this chapter, we've given you some starting points for your mixes and lots of advice on making plaster and applying it to your walls. In straw bale plastering, begin with the prep coats, then pack the walls — filling in holes between bales. When this is done, you

TRANSITIONS

As walls of a straw bale home take shape, they undergo a transition. The wall starts as straw. The next layer is a clay slip, which keys nicely into the roughened surface of the straw bale. Next comes the high-clay content discovery coat. Then comes the infill coat, with a slightly higher sand and straw content. The last coat, the finish coat, transitions back the other way — that is, toward more clay and less sand, in order to seal the wall.

can apply the discovery and infill coats. Next comes the finish coat. Be sure to key all plaster layers into the previous ones. Wet the walls as we've suggested.

We can't give you recipes for making plaster, because soils vary so much. Because of this, you will need to create your own mix. As you go along, you will find the right mixes for your site and will develop techniques that work well for you. Well, that's about it. You've seen how plasters are made and applied to walls. It's time to get out there and start mudding.

CHAPTER 7

Natural Finishes:
*Alises, Litema, Decorative Clay Finishes,
and Natural Paints*

NATURAL PLASTERING is a flexible and highly creative art. With a little practice, you can produce natural plaster walls with soft, delightfully inviting, and comforting lines. A good part of the artistry comes in the finish coat, which provides an opportunity to achieve the color and texture you'd like in your walls. Natural finishes, the subject of this chapter, also provide room for artistic expression (see color gallery).

Natural finishes are applied over the final coat of plaster. We'll examine four natural finishes: alises, litema, decorative clay finishes, and various natural paints. We'll look at how they are made and applied, and describe the uses of each. Before we get into them, however, a few words on pigments.

Understanding Pigments

Natural pigments are often used to provide color in a variety of ways in natural building. For example, they're used to color finish plasters and to tint alises, and also when making decorative clay finishes and various natural paints and limewashes. Before we examine these finishes, let's explore the fascinating world of pigments.

Humans have been using pigments to produce lasting colors for hundreds of thousands of years. The first known examples occur in the Paleolithic period, approximately 350,000 BC, at which time, studies show, our early ancestors used red earth pigments to tattoo their flesh and to decorate the bones of the dead. In the Middle Paleolithic period, about 40,000 BC, yellow ocher was used. Paintings on the walls of caves in France, made about 15,000 BC, contain brown, white, and yellow pigments.

Pigments consist of fine powders whose particles are suspended in some medium, typically water, and generally fall within one of two broad categories: natural and synthetic. Natural pigments come from three sources: earth, animals, and plants. Synthetic pigments, of course, are artificially produced, although some are chemically identical to natural pigments.

Natural Pigments

Among natural builders, by far the most widely used natural pigments are those derived from the Earth, notably colored rocks and colored soils (Figure 7-1). These materials are crushed (in the case of rock) or sieved (in the case of soil) to produce fine powders that are added to finish coats, alises, and natural paints. Some earth-derived pigments, such as the siennas, can be used directly; others are derived by heating a mineral pigment to intensify its color; for example, burnt sienna is produced this way.

The color of natural pigments comes from various minerals. Natural earth pigments frequently derive their color from the various chemical forms of the mineral iron, which is found in abundance. Different chemical forms of iron produce a stunning range of tones from red through yellow and green to violet.

One of the best known groups of earth pigments is known as the *ochers*, which contain a mixture of quartz sand, kaolin clay, and iron oxide and come in a wide range of hues from brown and yellow to red and violet. For example, ochers that contain hematite, a red iron oxide, may be red, red-orange, or even red-violet. Ochers containing yellow iron oxide are yellow, while ochers containing manganese oxide mixed with another type of mineral (geothite) produces the brown-toned pigments sienna and umber.

7-1: Gathering clay in the field can be fun and interesting. You'll find many different colors in nature that can be used for interior and exterior walls.

Natural earth pigments come in a variety of hues, some subtle, others quite vibrant (see color gallery). This enormous range of colors and hues offers the natural builder unrivaled opportunity for artistic outlet. Moreover, natural pigments tend to be extremely stable.

The Resource Guide at the end of the book lists suppliers who carry a rather remarkable selection of good quality natural pigments from around the world. Bear in mind, however, that even though they're natural, some pigments contain potentially harmful elements. Try to avoid those and always wear a certified dust mask when handling any powdered pigments or clays to prevent inhalation.

The color in a clay is due to the presence of various minerals. Colored clays are a good source of pigment for natural building, and can be harvested locally along roadways, hillsides, or in fields — especially in the western United States. We described ways to identify clay — and to distinguish it from silt — in Chapter 4. (Note: if you use powdered clay to pigment an alis or clay finish, you will probably need proportionately less clay in the base mix.)

Clays can be dug out and collected in buckets, then taken home and ground to form a powder. If you are collecting small amounts, you can grind them by hand, using a mortar and pestle. For larger quantities, you can screen the dirt or soak it in water, then screen the mix. Dry the pigments, then store them in air-tight, dry containers. Not all colored dirt you see along roadsides or in the painted deserts of the world are clays; they do, however, contain colorful mineral pigments that can be used as a source of pigment.

Colored clays can also be purchased from local pottery supply outlets as well as from various mail order outlets listed in our Resource Guide. However, pottery supply stores generally contain a limited number of colored clays, because most potters use glazes to achieve their final color and are generally not concerned with the color of the clay they use to make their pots.

On a final note, you will recall that natural pigments are also derived from plants and animals. As a rule, these are used for dyeing and staining materials, rather than for pigmenting finishes. Although there are some paints made from substances derived from plants, most of them are sensitive to light and therefore of limited value for painting walls.

Synthetic Pigments

Humans have been making their own pigments for thousands of years. The Egyptians, for instance, perfected the manufacture of blue pigments, partly to make up for a lack of naturally-occurring blue pigments, and used it to tint glass around the third millennium BC, as well as exporting it to other countries. These ingenious people also produced a darker artificial blue, actually a green tinted with cobalt. Over time, the Romans picked up on Egyptian techniques for manufacturing pigments and spread their knowledge and skills throughout their vast Empire.

Today, many pigments on the market are synthetic — that is, produced artificially. These products are fairly reliable (stable) and are more concentrated, rich, and vibrant than natural pigments, though they often lack the depth, subtlety, and beauty of earth pigments. Synthetic pigments are powdered and can be purchased in local pottery stores where they are sold as masonry pigments, or ordered from mail order supply outlets listed in the Resource Guide.

While synthetic pigments may have their place in society, they can also be toxic. Some pigments are made from toxic organic chemicals; others contain toxic additives. Still others contain toxic elements, such as cadmium, lead, or chromium. Although some

---

NATURAL EARTH PIGMENTS

Yellows: a variety of tones available; look for ochers and siennas

Reds: a relatively wide range of red pigments containing iron oxide; called Red Earths

Blues: limited

Green: pigments usually containing copper; referred to as Green Earths

Browns: numerous choices including raw and burnt umber

of these toxic elements can be found in natural pigments, they're usually found in higher concentrations in commercially available artificial pigments. Before buying natural or synthetic pigment, be sure to inquire about its toxicity.

Some Final Notes About Pigments

Working with pigments, both natural and synthetic, takes practice. Pigments vary in concentration: some are quite concentrated, and only small amounts are needed to achieve the desired color and hue. Others will require much greater amounts to achieve the color and hue you want.

When adding large quantities of pigment to an alis or decorative clay finishes, you may need to alter the mix a little — for example, use a little less clay to accommodate pigment containing clay. The key to pigmenting, as in making plasters, is to experiment. Prepare test samples, then apply them to your walls; let them dry and see how they look and feel. Be sure to record details of your recipe, the amounts of pigments you're adding and take detailed notes on the results.

Pigments purchased from suppliers are generally shipped in plastic or paper bags. When you receive a shipment, transfer the pigments to plastic screw top jars for storage. Remember, no matter what kind of pigment you are working with, wear a certified dust mask when handling or working with the material. Rubber gloves are also advisable. Work in a well-ventilated space, and keep all pigments out of the reach of children.

Alis: Natural Clay Finish

An alis is a thin earthen natural finish used for centuries in many different countries to protect and beautify earthen homes. In the desert Southwest, alises were traditionally applied by Native American women, known as the *enjarradora* (translated: plasterers), who were the undisputed experts on earthen building. The men prepared mud and transported it to walls, but it was the women who built, plastered, and maintained adobe buildings.

The final step in adobe construction was a "mud painting," or *alisando,* that was both decorative and functional. Using local clays, the Native American women applied the mud, or *alis* (*alisa* has been shortened to alis), to the walls using woolly (untanned) sheepskin to produce a suede-like texture (see Figure 7-2). Luminosity of the walls was often enhanced by the addition of mica.

Alises continue to be used today. Applied to earthen plaster finish coats, alises have the consistency of a heavy cream. When dried, they can produce stunning natural finishes that are warm and inviting.

Alises are usually applied with woolly sheepskin, sponges, and paint brushes and are used to enhance the beauty of interior walls. They are also used to provide a protective coat for outside walls as well as to refresh finish coats or older coats of alis.

Alises range from simple mixes of clay-rich soil to more complex mixtures containing manufactured powdered clay, fine sand, and flour paste added as a binding agent to create a harder, more durable and protective surface. Natural plasterers often add finely ground mica to their alises, to add hardness and produce a reflective quality that ranges from subtle to quite dazzling. Some natural plasterers add larger mica flakes. Still others add finely chopped straw, sage, leaves, or flower petals, which are just a few of the options.

Decorative alises are often made from white clay (kaolin), which provides a white base to which colorful pigments can be added. However, rich, beautiful colors can also be achieved by using colorful clays instead of kaolin. Newman Red clay, for instance, produces a rich orange/peach color, while Red Art clay produces a mauve color. Kaolin clay can be mixed with these clays to lighten them. (Experiment first to be sure your proportions are correct!)

Alises can be made many different ways with an endless number of variations of hue. We will provide you with guidelines that you can adjust to your particular needs.

7-2: Native American women of the desert Southwest were the undisputed experts on building. They applied alises to protect their dwellings using sheepskin as their applicators.

Benefits of Using an Alis

Alises provide a magical beauty that you must see to appreciate. Besides creating a beautiful natural finish, they protect interior and exterior walls, and can also refresh existing plaster finishes. If your finish plaster is a bit dusty, alises (containing flour or casein) can prevent dust from becoming a household nuisance. In addition, they provide a surface that can be refinished down the road when you are ready for a change.

Alises are extremely easy to prepare. They're typically made from a slip produced from local clay-rich dirt or from a mix of fine-grained sand, flour paste, pigment, and powdered clay. Once you've prepared a couple of batches of alis, you'll appreciate how simple alis-making really is. As you begin to develop a feel for the process, you can experiment with the ratios of ingre-

PROVIDING MORE THAN PROTECTION

Alises not only provide protection for the underlying plaster and wall, they provide an opportunity to achieve unrivaled beauty — adding subtle or rich color.

ST. FRANCIS' ANNUAL FACE LIFT

Each year, St. Francis de Assisi, a large adobe church located just south of town in Ranchos de Taos, New Mexico, gets a new coat of alis. Over a two-week period, the parishioners of this Catholic church join forces to revive their plaster finish by applying seven thin coats of a clay alis. They use a locally harvested high clay-and-silt soil containing minimal amounts of sand, and mix it with water to make a simple clay slip. Walls are wetted, then rubbed or brushed to prepare the surface — that is, to remove dust and loose plaster. Plaster repairs are made and then the volunteers use sheepskin to apply the alis. One layer of alis is applied for each major anticipated rain storm. Alis is anually applied to restore the beauty of the finish and to continue safeguarding plasters against the elements (see color gallery).

dients — as well as the type of ingredients — to produce the ideal material for your walls. You can even use alises as a household paint over drywall!

Besides being simple to make, alises are fairly inexpensive to produce and extremely easy to apply. Tool requirements are minimal, and clean up is a snap: water and a brush are all that's required. There's no need for potentially toxic solvents and no washing of potentially harmful chemicals down the drain.

One of the most important benefits of alises for natural builders is their breathability — or more correctly, their permeability to water vapor. Being made primarily of clay, alises allow water vapor to flow out of walls. This protects earthen and straw bale walls from the damaging effects of moisture, a topic we've covered in Chapters 2 and 4. Clay in alises also seals walls from moisture, as described in Chapters 2 and 4.

How Are Alises Made?

In this section, we'll examine two basic types of alis: protective exterior and decorative interior.

PROTECTIVE EXTERIOR ALISES. As their name implies, protective alises are used primarily for exterior applications. That is, they're applied over an earthen plaster finish coat, the final coat of plaster, where they provide an additional measure of protection. They can be applied soon after the final coat of plaster has dried or at a later date — for example, when the finish coat looks like it needs a "refresher." They can also be applied over plaster repairs.

Protective alises are typically made from the same soil one uses to create plaster for an earthen finish coat, and thus won't change the final color of the wall. However, protective alises can also be made from others soils, clays, or pigments, thus allowing an applicator to change the wall color. The problem with this approach is that, as the protective layer on exterior walls wears away, the base plaster color will begin to show through. (This is why most people use the same soil for their exterior alis finish as for the final coat of plaster.)

To make a protective exterior alis, begin by screening the clay-rich soil, as you would for any plaster. Mix the screened soil with water to create a creamy slip (having the consistency of heavy cream). Be sure the sand and aggregate sink to the bottom before extracting slip. This slip is now ready to apply using a brush or sheepskin. Details on the application of aliases are presented below.

In sum, exterior alis finishes provide extra protective coats to walls and are also used to refresh and maintain exterior walls. The climate, design details, severity of weather, and other factors determine how many layers are necessary, and frequency of application.

DECORATIVE INTERIOR ALISES. Decorative interior aliases are typically applied in two coats — each prepared slightly differently. Both are made from bagged powdered clay or clay extracted from subsoil, fine sand, water, and pigments. Mica and flour paste are often added to them as well. Flour paste makes the mix more adhesive and makes the alis more durable after drying. The second coat of alis typically contains a finer-grained sand and can contain decorative elements such as large mica flakes, sage, and finely chopped straw. Flour paste or casein in sufficient quantity reduces dusting of this layer.

SOURCES OF CLAY. Alis can be made from local clay-rich dirt, or from bagged, powdered clay available from pottery supply stores. Using local clay-dirt to make alis is more labor intensive than using powdered clay, but it's cheaper and requires less energy.

For decorative clay finishes, many plasterers begin with kaolin clay, a white powdered clay. Kaolin clay can be purchased in 50-pound (about 23 kilogram) bags at local pottery supply stores or sources listed in the Resource Guide — and is usually quite inexpensive. Other types of powdered clay work well, too, but as mentioned earlier, white clay gives you a nice starting point for pigmenting: it is easier to achieve a desired color when your base material is white. Be aware, however, that although all "white" clays, including kaolin, appear white, they usually have some other tint (for example, a slight gray tint), which may alter the final color of your mix and thus confound your efforts to achieve the color you're after. A white clay with a slight gray tone, for instance, may produce a colder, duller hue than you'd like.

> ### TOOLS AND MATERIALS NEEDED TO MAKE AND APPLY AN ALIS
>
> - Clay dirt or powdered clay
> - Fine-grained sand
> - Cooked flour paste
> - Water
> - Wheelbarrow or cement mixer (or some other container to mix it in)
> - Window screen
> - Buckets
> - Paddle mixer and electric drill
> - Sheepskin, paint brush, or sponge for application
> - Sponge and disks for buffing
> - Sponge and rags for clean up

7-3: These "paint pots" found in Yellowstone Park are one of many natural sources of kaolin clay.

CEDAR ROSE GUELBERTH

SAND IN ALISES. Clean, fine-textured sand is used in decorative alises. A good choice is silica sand (40 – 90 grit) which is available at building supply outlets. Silica sands are recommended in part because they are usually light-colored; other sands are often dark or may contain dirt that can alter the color of an alis. Remember: the finer the sand, the smoother the finish. Use 40 - 60 grit for the first coat and 60 - 90 grit (a finer sand) for the second coat of alis. Adding mica produces a smoother coat.

ADDING MICA TO ALIS. Mica is a naturally occurring mineral (chemically similar to clay) which forms large flakes that can be ground up to produce soft, fine flakes or powder — both of which have a sensual feel. (Suppliers are listed in the Resource Guide.) When added to finish plasters (color coats) and alises, mica produces a hard and more durable finish. Fine mica flakes also provide sparkle and sheen. (Note: the finer the mica, the less sparkle. However, finer mica provides hardness and a reflective quality that provides more depth to your finish.)

Mica has been used traditionally in finishes throughout the world where available locally. As noted earlier, native American women of the desert Southwest used mica in their alises.

Few speak more glowingly of the use of mica than Carole Crews of Taos, New Mexico. "If it is easily available," says Carole, "mica makes a lovely addition to finish plasters as well as clay slips and adds to their workability." Besides adding beauty to the wall, mica makes the alis go on easier. She goes on to explain, "[Mica] is like a molecularly flat sand, which is smooth instead of gritty." (See Figure 7-5.)

Purchased in bag form, mica comes in range of grain sizes: from 260 grit powder to fine (1 - 2 mm) flakes. Larger flakes can be added to enhance the aesthetics of the wall. (Note: the surface of the wall must be buffed to bring out the mica.) When using mica, bear in mind that you may need to reduce the amount of silica sand in your mix.

FLOUR PASTE. Flour paste can also be added to an alis mix. It serves several useful purposes — for example, it makes an alis smoother and easier to apply and also makes it adhere well to an earthen finish plaster and other surfaces. Flour paste creates a more durable alis and prevents dusting. (The recipe for making flour paste is in Chapter 6 on page 114.) As a reminder, be sure that flour paste does not burn (which discolors the material) and that it is free of lumps. Clean out the pot from one batch to the next to prevent burning.

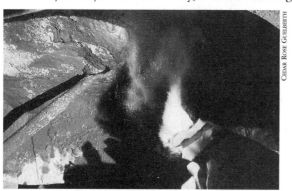

7-4: Adding mica to a plaster mix, as shown here, makes the walls sparkle and increases the durability of the your surface.

CEDAR ROSE GUELBERTH

7-5: Carole Crews has been a pioneer in the development of beautiful alis finishes, transforming the traditional techniques to produce spectacular finishes.

CEDAR ROSE GUELBERTH

Making an Alis

Although the alises we'll describe here are made from flour paste, clay, fine sand, and mica, other ingredients — or combinations of ingredients — can be used. These include casein (milk protein), milk, lime putty, and quartz dust. Finely chopped straw and leaves can be added to the final alis coat for decorative purposes.

Small batches can be mixed in one- to five-gallon buckets. Larger batches can be mixed in a wheelbarrow, plaster or mortar boat, or cement mixer. (Be sure your tools are well cleaned and free of other materials that could discolor your mix.)

You may want to start out by making a small amount — a quart or a gallon — of decorative interior alis, just to get the hang of it. Begin with this ratio: two parts kaolin clay to two parts fine silica sand (or mica-sand combination) to one part flour paste. Be sure to wear a certified dust mask to prevent inhalation of fine particles, especially when gearing up for mass production.

To mix your alis, begin by adding sand, mica, and clay to flour paste in a clean container. Mix well to eliminate clumps. Add water as needed until the mix attains the consistency of sour cream or a thick milk shake. Resist the temptation to dilute the mix past this point: it will go on too thin if you do — and will drip all over the place! Stir small batches with a whisk.

When making large quantities of alis for big jobs, you can use a large plastic tub (Figure 7-6) or a cement mixer. To mix in a five-gallon bucket, begin by adding flour paste — about ⅕ of a bucket. Add one quart of fine sand and one to two quarts of mica powder. Stir the mix, then add three quarts of kaolin clay. After adding the sand, you can stir by hand using a large whisk. Or, you can use a paddle mixer attached to an electric drill. The second coat of alis is made similarly, except finer sand is used.

ADDING PIGMENTS TO AN ALIS. Once the alis base is thoroughly mixed, it is time to add the pigment. Pigmenting is relatively easy step. As noted earlier in the chapter, there are several sources of pigment. There are several ways of adding a pigment to an alis base mix. Powdered pigments can be added directly to the base mix. Add a small amount at a time while mixing to ensure even distribution of the color.

MAKING DECORATIVE ALIS FROM SOIL

Alis can be made from soil, preferably dirt with a sufficient amount of clay. Sift the dirt through a 1/8-inch hardware cloth. You may need to add sand, depending on the amount of sand and silt in your dirt. Add flour paste and mica, if required. Experiment with pigments if you want to alter the natural color of the dirt. Alis can also be made from clay slip. Flour paste can be added to increase its durability. Pigment can be added for color.

CAUTION

Be sure to wear a certified dust mask to prevent inhalation of the silica, mica sand, clay, and pigment. Sustained inhalation of the siliceous dust can lead to silicosis, a debilitating lung disease.

7-6: Alis for large jobs can be mixed in large tubs like this one.

CEDAR ROSE GUELBERTH

Powdered pigments can also be mixed with warm water, and then added to an already-mixed alis base. Place the pigment in a jar of warm water, then shake. When the pigment is thoroughly suspended in solution, usually after a minute or two of agitation, pour it into the alis base. Stir the mix until the color is evenly dispersed. (Be sure to compensate for the extra water in your mix.)

Pigments are also available in wet form — that is, already mixed with water — to use in making the alis. However, be aware: pigments may be dissolved in chemicals you don't want in your plaster.

TEST COLOR PATCHES. Dry pigments come in a wide variety of colors. Be aware that when added to a wet mix, then dried, the final color may differ considerably from the color of the original pigment. Therefore it is important to test different pigments, and the best way to do this is to start with a kaolin base alis. Divide the mix up and place in smaller containers, then add a different pigment color to each container, and apply test patches to the wall. Mark each patch clearly and record what colors were used in each patch (Figure 7-7).

7-7: Color test patches on a wall allow one to determine the alis or colored clay plaster mix that provides just the right color.

CEDAR ROSE GUELBERTH

Once you determine a suitable color, you will need to experiment with different concentrations of the pigment to yield the depth of color you prefer. Be sure to use the same amount of base in each sample and keep track of the amount of pigment you add to each sample. This information will allow you to determine the amount of pigment you need to add to your base mix. (Remember: you may need to alter the clay content of your base mix when using a pigment with high clay content.)

We recommend creating several test patches. (See color gallery) Mist the wall surface and then apply test patches. Examine how each one goes on the wall and record your observations. After the alis has set up, you may want to rub each test patch with a clean damp sponge or a plastic disk using circular motions. (This polishing is an optional step. Details on this process are presented shortly.) Then let them dry.

After the test patches have thoroughly dried, check to see which, if any, is the color you want. As you will soon discover, alises are lighter in color when dry than when they are wet. In fact, some pigments vary in color and depth of color tremendously from wet to dry. (That's why test patches are so important!)

If the alises are too light, add more pigment. If they are too rich in color, add less pigment. (You can dilute your test mix by adding more white base mix to achieve the color you want.)

Once you've obtained the color you want, it is important to examine the quality of the dried alis coat. How would you describe its texture? Is it smooth or grainy? If it is grainy, that's a sign that you've probably added too much sand — or that you may need a finer sand. Does it dust off? If it does, it's a sign that your alis contains too much sand or not enough flour paste or clay. If the alis did not provide even coverage and seems to be thin in places, you need to add less water to the mix. If it is "lumpy" and doesn't spread well, it probably needs to be mixed more thoroughly. It may also require a little more water (not too much!) or you may need to screen the lumps out of the flour paste.

Adjust your mix accordingly and apply another test patch or two, then polish it and wait for it to dry. Assess your results. Continue this process until you produce the finish that pleases you — that is, until you've reached the desired color, texture, and durability. As always, be sure to take notes. It's amazing how quickly you can forget details when you're running a number of tests.

Additives

A simple decorative alis made from sand, clay, mica powder, flour paste and water should suffice for virtually all applications. But like earthen plasters, alises are often supplemented by adding various substances. Some of these additives enhance the performance of an alis; others enhance its beauty.

Powdered milk (one cup to three gallons) is sometimes added to an alis — alone or in combination with flour paste. Powdered milk thickens the mixture and also acts as a binder. An alis to which powdered milk has been added tends to be a little harder, less likely to dust, and more durable than one made without additives.

To alter the aesthetics, you can be very creative. Some natural plasterers like to add a little finely chopped straw (one- to two-inch lengths) to alises. A couple of handfuls in a five-gallon bucket works fine. You'll have to decide for yourself how much you want to add.

Flake mica, ranging from 1/8 to 1/4 inch, can also be added to the final coat of alis. Larger pieces can be applied to the wall while still wet. Sage, dried flowers, or leaves can also be added to the mix, or applied by hand to the wall. Be creative, experiment, have fun. But remember that when you add mica flakes or other decorative items to your alis you will need to buff and polish the surface so you can see them!

TESTING COLORS

To test colors, make up a batch of kaolin, sand, cooked flour paste, and mica (if you want to include it). To experiment with different concentrations of color and/or different pigments, begin by taking one cup of the mix and add pigment. Mix well. Then place another cup in a separate container and add a different pigment — or more of the same to achieve a darker alis. Mix it. Repeat with as many samples as you want, and make notes about each. Apply the alises to a lightly misted wall, then let them dry. Buff the surface, if desired, then select the pigments or concentrations you like the best. Be sure to save a little of each test batch for color comparison — that is, to ensure you get the right tint when you're mixing larger batches. Mark each wall sample well for later reference. (See color gallery.)

How Are Alises Applied? Alises were traditionally applied by pueblo dwellers with a sheepskin (a fleece, not a tanned skin) as part of the annual spring cleaning and maintenance of their adobe buildings — and is still applied this way today (see Figure 7-8). Alises are applied in thin layers to restore the wall finish, create a protective coating, and, in the interiors, to cover up soot that accumulates on the walls over the winter.

You can use sheep's wool to apply your alis, although it takes practice to master this technique so that the alis is applied evenly. Most natural builders use three- or four-inch wide paint brushes. Smaller one-inch brushes may be used for tighter spots — along ceilings or around window trim. A soft-bristled brush will gum up and a hard, stiff-bristled brush will not hold alis well and could cause streaking. Use a brush that will hold the material but causes minimal streaking. You may have to try several types to find the right one.

CEDAR ROSE GUELBERTH

You can also apply alis with a palette knife along trim or along the base of walls. Eliminating a little water in the mix will thicken it a bit and facilitate application. Alternatively, along the base of a wall, you may want to use a sponge to apply alis. Dip it in the alis and slide the sponge upward for best coverage. Small brushes and small flexible trowels work well here, too. Thicker alises can be applied by hand then troweled smooth. Alis goes on very quickly this way.

Before you can apply an alis, however, it is important that an earthen plaster finish coat looks exactly the way you want it to look. (In other words, if you want a smooth finish, be sure the finish coat is smooth. Alis is not thick enough to fill in dips or unevenness or rough texture.)

Once you have prepared your alis, pour it into a small bucket. Tape off any wood, as alis may stain wood, especially rough-cut lumber. Although alis will come off smooth, sanded wood if you wipe it up pretty quickly with a clean, damp rag, it is still a good idea to protect wood from potential staining. Mask to get clean edges or lines, and lay drop cloths down to avoid splattering the floor, then begin applying the alis, working from the top of the wall down to avoid drips and runs.

7-8: Alises may be applied in a number of ways. Here a woman applies alis the traditional way, with a sheepskin.

Before applying an alis, lightly mist the finish plaster. In most instances, two coats of alis are applied to a wall. As the first coat begins to dry, the wall will begin to appear mottled and will start to take on the hardness of leather. (Mottling begins about an hour after application, but drying time depends on weather and ventilation — see Figure 7-9.)

As the wall becomes mottled, you can polish the alis to remove brush strokes. As described below, polishing results in a smooth, finely textured surface.

After the first coat of alis is completely dry, it is time to apply the second and, we hope, final coat. As noted earlier, this one usually contains finer grain sand or, in some cases, more mica and less sand. Paint it on the wall as you did the first coat. Apply leaves or large mica pieces, if desired. Once it sets up, but before it is completely dry, you can buff or polish the surface with the plastic lid of a yogurt container. Once an alis has completely dried, you can polish it with a sponge to bring out the mica or other decorative elements.

CEDAR ROSE GUELBERTH

7-9: As an alis or natural clay color coat sets up and begin to dry, it looks blotchy, as shown here. At this stage, the wall can be buffed. When it dries completely, the blotchiness disappears, creating a consistent, beautiful finish.

To polish a wall, dip a sponge into a bucket of warm water, then squeeze out the water really well: you want a slightly damp sponge. Gently rub the wall in circular motions (see Figure 6-18). Be sure to dip the sponge in the water and rinse out frequently as you polish the wall. If the sponge starts to accumulate sand particles it will become abrasive and will begin to scour the alis, producing a rough surface. As you polish the wall, loose bits of mica and straw (if you've added some to your alis) will flake off. Polishing should bring out the beauty of the wall — showing off the mica, straw, or other decorative elements. Polishing results in a smooth, finely textured surface.

SOME FINAL POINTERS ON ALIS. Now that we've shown you how to make and apply an alis to a plastered wall, we want to share a few useful pointers. First, remember that alises containing flour paste and straw need to be applied fairly soon after they're made, though they can be stored in a refrigerator for several days. If an alis goes bad on you, discard it: the unpleasant (sour) smell will not go away, even after the alis has dried on a wall.

Second, if you have excess alis, don't pitch it: you may need it later. But rather than store it wet, dry it out. To dry, pour the alis onto clean plastic sheeting or baking pans in an area where it won't be contaminated by blowing dust, debris, and straw; then let it dry

DAN CHIRAS

7-10: When the alis in these cake pans has dried, it can be stored in plastic bags for later use.

(Figure 7-10). After the water has evaporated, cut the alis into cookie-sized chunks and store in an air-tight dry container. Dried alis containing flour paste and straw can be stored in waterproof containers for long periods for later use — for example, when you want to refinish the wall or repair a section. When that time comes, you simply rehydrate the mix and apply it as before. (Be sure to save some dry pigment, too, as you may need it to make more alis to match your original.)

In closing, decorative alis applications work very well for interior walls. On exterior walls, however, alises are exposed to elements and are therefore rather vulnerable. Exterior alis provides protection, but will eventually wear off if exposed directly to the elements, especially rain. The more a design protects exterior walls, the longer an alis will last. As in a conventional painted house, you will occasionally need to re-apply alis to maintain a superior finish.

Litema: Clay-Dung Plaster

In India and South Africa, many earthen homes are painted with brightly colored clay plaster (see color gallery). Made from dung and colored clays, this plaster, called "litema" (pronounced dee-TAY-ma), is a decorative and quite protective wall finish.

The folks at the Cob Cottage Company in Oregon use it, especially "on areas that are likely to get scraped or knocked, such as around windows and doors," according to Michael Smith, author of *The Cobber's Companion*. Smith explains that litema "has a pliable, fibrous texture and is moderately resistant to rain. It dries with no odor, so it can be used either inside or out."

In India and South Africa, fresh cow manure is mixed in equal amounts with brightly colored clays and pigments. If fresh or slightly aged horse manure is used, the mix of dung and clay is left for a few days to ferment. During this time, the dung binds to the clay, creating a durable plaster. After fermenting, the mix is smeared on walls.

Mist the walls first, then apply a light coat by hand. If you're using fresh manure, the mix will feel greasy and will slide on the wall with ease. Several coats can be applied to create a durable finish.

Decorative Clay Finish Coats

Decorative clay finish coats are yet another option for finishing your walls. They're very similar to alises, but made into a thicker, more pliable mix, and are applied using a trowel, rather than a paint brush. Decorative clay finish is a two-coat color finish plaster applied over an earthen finish coat or a smooth, earthen plaster infill coat (with a slightly rough texture so the clay finish coat keys in well). Like alises, they allow for lots of color options, including pure white. Applicators can achieve a variety of textures from a rough look with trowel marks to a buffed, smooth surface (see color gallery).

Made from ingredients similar to decorative alis finishes — notably kaolin clay, silica sand, flour paste, and fine mica — decorative clay finishes are applied in thin plaster coats approximately ⅛-inch thick. For the first coat, try three parts silica sand (40-60 grit), one part fine mica flakes or powder, three parts kaolin clay, one part screened flour paste, and pigments. Add water as needed to produce a mix with a whipped cream cheese consistency.

For the second coat, use a finer-grain sand, for example, silica sand (60 - 90 grit). You can also increase the mica content (while decreasing the sand), which increases the reflective quality and hardness of the surface. In addition, you may want to slightly increase the amount of flour paste you add to produce a more durable, dust-free surface.

As in plasters and other finishes, experiment with the ratio of the ingredients to arrive at the best mix. Remember, that you may need to compensate for the clay content of the pigments you use. The mix is drier than an alis, should be pliable, and easy to apply evenly and consistently with a trowel (Figure 2.6).

Be sure the wall surface is exactly as you want it prior to applying this finish coat. Although this finish can be used to fill in slight dips and smooth out your surface, don't count on it for significant infilling. Because of this, decorative clay finish coats are best applied over a fairly smooth finish coat. However, if you even out your infill coat fairly well, you can use this material for your final application.

To begin, moisten the wall. Trowel on the first coat about an ⅛-inch thick. After this coat has dried leather-hard, smooth out trowel lines, if you want. Once this coat is completely dry, moisten the wall, then apply the final coat, again ⅛-inch thick. When applying the final coat, be sure you can finish a complete wall at one time to eliminate cold seams. A cold seam is an area on which the plaster dries before an adjoining section is applied, for example, when one day's work ends and another begins — or even when a lunch break interrupts application.

The final coat can be left to dry with trowel marks or can be buffed when leather hard with a plastic lid or a slightly moist sponge, or hard troweled to create a smooth surface (see Figure 6-18). If you buff or trowel the finish coat too soon, however, you will very likely smear the finish and cause cracking. Be sure to buff in circular motions and from different directions. Use a damp sponge, plastic lid, or hard trowel. If the finish cracks, you may have to adjust the mix, adding more sand. Cracking may also be caused by buffing the wall too early or "overbuffing" the finish. If the finish is too hard and you cannot smooth it out, you buffed too late.

Decorative finish coats can also be made from clay-dirt, rather than powdered clay and mica, as noted earlier. If you elect to make a finish coat from clay-dirt, you will have to experiment with your materials to arrive at the proper mix. Be sure to sift your dirt through a metal window screen and use fine silica sand, fine mica, and screened flour paste.

As in any plastering or finishing project, be sure to apply several test patches to determine color, texture, and finish quality (see color gallery). Let them dry and assess their appearance and performance. If the finish dusts, add more flour paste. If the color needs adjusting do so. If it peels off the wall, make sure you have wetted the wall well and that the mix is wet enough. If the mix doesn't stay on your trowel as you are apply it, it may not be thick enough.

Have fun experimenting. You'll be glad you did.

Protecting Earthen Plasters with Silica Paint

Earthen and lime plasters can also be finished with silica paint (also known as *waterglass*), which is translucent. Made from potassium silicate and pigment, silica paint can be used for interior and some exterior applications.

SILICA PAINT FOR EARTHEN AND LIME PLASTERS

Silica paint is good for earthen and lime plasters. When applied to lime plasters, it creates a fairly water-resistant finish. So if you need to protect a lime-plastered wall, say an exterior wall or a bathroom wall from moisture, silica paint may be the cure. Applied to earthen plaster, it produces a clear, durable finish; however, silica paint does not produce a very water-resistant surface in earthen plasters.

Silica paints create a hard, protective finish that prevents dusting. If your finish coat is dusting, a silica paint will solve the problem. If you want to preserve the color of the earthen finish coat, omit the pigment from the silica paint.

To make a silica paint, start with a liquid potassium silicate solution. Because it is available in different dilutions, you'll need to experiment a bit. Start with an undiluted solution, then add pigment. Apply a little on a wall. If it goes on too thickly, dilute it with some water. Try again and dilute it till you find a solution that works well — that is, it goes on the wall smoothly, penetrates the surface, and produces the look you want. You can also add part quartz dust to the potassium silicate solution to produce a more opaque paint. (Potassium silicate can be purchased from BFH, a company listed in the Resource Guide.)

Unfortunately, not all pigments work with silica paints: some pigments undergo chemical reaction and change colors dramatically. Some fade fast; some change color over time; some clump in solution and won't mix in. So be sure to experiment a bit first. For best results, use only natural mineral or earth pigments and avoid chrome, lead, and cobalt pigments.

Once you've found a solution that works and the solution is thoroughly mixed, you'll need to prime the wall with a dilute solution of unpigmented potassium silicate. Dilute your solution by adding one part waterglass to five parts water. If the "primer" is quickly absorbed into the wall, you know you've got the correct solution. Be sure to add silica sand to your finish plaster for best adhesion of a silica paint.

Next, apply the paint using a brush. Once it has dried, you can apply additional layers. When they are dry, you can seal the paint by applying an overcoat. Use the diluted solution of unpigmented potassium silicate.

As a final note, remember: once you've made silica paint, store it in plastic containers, as potassium silicate can etch glass. When applying silca paint, be sure to cover windows and metal, too, as the solution will corrode metal.

Casein Paints

Throughout this book, we've mentioned casein and milk paints, which are a type of natural paint, for application to earthen plaster. What is it? Where do you get it? How do you make it?

Casein paints were popular in Europe and other countries for thousands of years, but like many other natural materials, they have been slowly squeezed out by modern products, in this case by commercial oil and latex paints. Today, milk paints are used to create beautiful, safe surfaces with Old World depth and charm. (See color gallery)

Casein paint is paint made the traditional way — that is, from milk. Casein is a protein in milk, which when extracted and dried, can be mixed with water to produce a translucent finish. When applied over an earthen finish coat, it allows the natural color of the wall to show through. (Translucent casein paint may also be applied to reduce or eliminate dusting of a finish coat.)

Casein is also mixed with pigments to produce a paint that is applied to all kinds of walls, including those plastered with earth. Chalk, lime, clays, and pigments such as titanium white can be mixed into the casein to create a white base to which other colors can be added or pigment can be added to a translucent base to create a glaze type finish.

You can make your own casein paint or purchase it in powdered form from several retailers listed in the Resource Guide. Prepackaged casein paints typically contain casein, lime, clay, cellulose, asbestos-free talc, sodium phosphate, and pigments.

The renewed interest in casein paint for walls is a result of our changing aesthetics. In recent times, the aim of paint has been to make our surroundings bright, shiny, and washable. Today, however, with the resurgence of natural building, more and more people are seeking a different look — a more natural or organic feel for their homes. The rich, flat texture of casein paints produces a simpler, softer finish, especially when used in conjunction with natural building materials such as earthen plaster. Natural home builders are also looking for materials that produce a vapor-permeable (breathable) surface. Casein fits the bill nicely.

CASEIN PAINTS

Prepackaged casein paints come in powder form for long shelf life and less energy use and cost in transport. They are frequently suitable for people with sensitivities to chemicals. If that is a concern for you, request a sample to test its effect before you apply casein paints as some people react to these paints (Figure 7-11).

Current interest in casein and milk paints also stems from our awareness of our limited natural resources and other environmental concerns, especially concerns for indoor air quality and human health. These concerns have directed our gaze towards natural, biodegradable, nontoxic, natural, and solvent-free paints, stains, and finishes.

Before we get into the details of making casein paints, a few words on this material. First, remember that casein is a time-tested product: it creates a durable finish that has been used for centuries with great success. Second, casein paint is not waterproof and therefore should only be used in interior applications. Third, several coats may be required for full coverage. Fourth, casein paint may crack if applied too thickly over plastered surfaces. Fifth, although casein paint is not waterproof, it is very durable and does not rub off walls and can be lightly washed. Sixth, casein is considered hypo-allergenic, which means that it is less likely than many other paint products to cause allergic responses in individuals. Note, however, that many casein paints have a slight odor that may be a problem for some sensitive people: test first. Seventh, casein paint is easy to touch up. When you've finished a job, dry out a little of your mix and save it in case you need it later for touch ups. Eighth, as noted above, casein paint is breathable and is an excellent finish for natural building where breathability is paramount. Ninth, casein paint generally produces a flat finish.

7-11: Prepackaged casein paints like these come in powder form. Mix them with water and apply

Be aware: milk protein in the paint can support microorganisms, such as bacteria, mold, and fungi, that may cause the paint to spoil. (This occurs after water has been added.) Lime or borax may be added to casein paints to prevent growth of bacteria and mold. If not, the paint will sour in a few days. Don't apply the paint if it smells bad, as the odor will persist long after it is applied. As a general rule, you should not store liquid casein paint — that is, powder mixed with water — for more than a couple of days. If you must store it, keep it in a refrigerator to retard microbial decomposition. (If it contains borax or lime, it could last two weeks before going sour on you.) To save for a longer period, allow it to dry out completely, then store it in an air-tight container. Rehydrate it when you are ready to use it. Although casein paint doesn't store well after being mixed with water, casein powder or prepackaged casein paints can be stored safely for long periods in a dry, cool environment.

Casein paints should not be applied in areas where moisture is a problem — for example, homes with high levels of internal humidity (Earthships) or in bathrooms and possibly utility rooms and kitchens — unless they're well ventilated and lime or borax has been added to the paint.

Purchasing Casein Paint

The easiest way to obtain casein paint is to order the powder from a manufacturer such as Bioshield or the Old Fashioned Milk Paint Company, a family-owned business in Groton, Massachusetts. You can also purchase these products through various retailers listed in our Resource Guide. Casein paint from Bioshield comes as a white powder that is colored by adding pigments. The Old Fashioned Milk Paint Company paints come in sixteen colors as well as a translucent base.

Making Your Own Milk and Casein Paint

Milk-protein paints can be made in one of two ways, from skim milk or from dry casein powder purchased from various suppliers listed in the Resource Guide. Let's begin with skim milk

MILK PAINT. For those who want to make their own milk paints, here's a recipe you can try. Begin by taking a quart of low-fat or skim milk. (We're starting small until you get the hang of it.) Add two teaspoons of low-fat sour cream. Whip well, then let it sit in a warm spot until the milk has thickened. Once the milk has thickened, warm it on a stove or add a small amount of lemon juice or vinegar: this causes the mix to curdle, that is, to become lumpy. Curdling may occur overnight or may take a day or two until it separates into a solid material, known as *curd*, and a liquid called *whey*. Curd is coagulated milk protein: it's the stuff from which we'll be making our paint. Note: You can also create curd by simply adding lemon juice to the milk. (In other words, you don't need to add the sour cream). The acid in lemon juice causes the milk protein to coagulate, thus forming chunks called *curd*.

Next, separate the curd from the whey using a cheese cloth filter — for example, pour the curdled milk through cheese cloth placed in a colander over a sink. Wash the curd in the cheese cloth with clean water. Curd can be used as is or can be dried to produce a casein power.

The curd is blended to make a thin, translucent base. Chalk, lime, clay, or titanium white can be added to produce a white base. Pigments can be added as well for color. Because milk paint is thin, several coats will be required to cover a surface.

CASEIN PAINT. Casein paint is made from powdered casein, which can be purchased from vendors. The result is a rich and creamy, full-bodied paint — considerably thicker than milk paint produced by the recipe we just gave you.

Here's how you make it: to begin, place about four ounces of powdered casein in a glass or enamel container (not a metal one), containing 16 ounces of distilled water. Mix and let it stand for a few hours, stirring occasionally. The result is a casein slurry.

Next, add two ounces of powdered borax to 16 ounces of distilled water *in a separate container*. Then add the borax solution to the casein slurry and stir well. Wait several hours until chunks or particles of casein have disappeared.

Heat the mix in a double boiler to 140° – 160° F (60 – 70° C) When it turns into a relatively translucent liquid, remove from the stove and let it cool. What you've got at this point is a "casein wash." This translucent base can be used as a primer or a clear coat. You can also add chalk, clay, titanium white, or lime to create a white base to prime the surface of a wall — for example, an earthen or gypsum plaster or drywall. Remove some for use as a primer. You will use the rest of the casein base to make paint.

To make a paint from the casein base, start with some powdered chalk or clay mixed with water. Start small, say a cup or two to begin. Mix the chalk, clay, or titanium white in water to form a thick paste. (Suppliers of powdered chalk and clay are listed in the Resource Guide.) Then add the casein base. Because casein is an extremely strong binder, you will need to dilute the casein base before adding it to clay and pigments. However, too much water will make it thin and it won't cover well. Next add titanium white to create a white base and/or pigments for color.

Try a little of your paint out on a piece of cardboard. If it rubs off easily, you need to add more casein base or dilute your casein less. If it cracks, dilute the casein solution some more and add more chalk/clay/pigment. You can refrigerate home-made milk paint for two to three weeks, so long as you've added borax. Be sure to warm it up to room temperature before you apply it, however.

Applying Casein and Milk Paints

Casein and milk paints are applied by brush or roller: paint as you would any wall. A mixture of (unpigmented) casein base and water is a good primer, or you can prime with a white casein base coat. Because casein paints tend to produce uneven coverage, two coats may be necessary to coat a wall thoroughly.

If you haven't applied a primer coat, the paint may be absorbed by the wall surface very quickly and unevenly. A primer coat evens out the porosity of a wall surface, producing a uniform base to ensure even coverage, and acts as a binder between the wall and the next layer of paint.

Always run tests. Prime a section first; when it dries, paint it. If the paint dusts off the wall (comes off when you rub it), you didn't prime the wall adequately or you may not have put enough casein in the base mix. Try again. However, don't go overboard: too much pure casein base primer can create a glassy surface that will cause subsequent coats of casein paint to flake off.

Some people add oil to their casein paints to create a more waterproof coating. Linseed oil works well. Boiled linseed oil dries faster than unboiled oil. Add a little water to linseed oil, then place it in the sun, which causes it to thicken, then whip it into casein paint. Tung oil and poppy seed oil work well, too.

Oil can be added to the washed curds then whipped. Oil should constitute about 25% of the volume of the paint. (You'll need to experiment a bit with your mix. If it dries too slowly, you may need to reduce the amount of oil you add.) Pigment can be added to the casein-oil emulsion as well.

For best results, use a white casein base coat, followed by one or more thin casein-oil pigmented coats — called *glaze coats*. Let each coat dry before applying a new one. Beautiful effects can be achieved by painting one color over another.

EGG PAINTS. Casein paints are not the only options for natural building, but one of the most popular and convenient. Paints can also be made from pigments mixed with other binders, such as egg yolk or egg white.

Egg yolk paints have a long history. They were used by Flemish artist and painters in Renaissance Italy, and have also been applied to walls and furniture.

TIPS ON APPLYING CASEIN PAINT OVER PAINTED SURFACES

When applying casein paint to a previously painted surface, the casein paint may crack. This may be a result of a poorly made paint, which means you need to adjust your formula. The cracks may also be caused by an incompatibility between the casein paint and the previous coat of paint. To prevent this from happening, you can prime the wall with a bonding agent — either a casein base or Old Fashioned Milk Paint Extra Bond. Then apply your color coats.

DURABILITY

To produce a more durable paint, casein powder (which is acidic) can be mixed with an alkali or base, such as lime putty, soda, or borax. These additives deter mold growth. One part casein to four parts lime putty produces an excellent, durable paint. Such paints have been used for many centuries to paint furniture and walls.

Egg yolk paints are made by separating egg yolks from the white, then piercing and removing the membrane that surrounds the yolk. Pigment is added next, a little at a time — about 1/4 teaspoon to 1 teaspoon per egg yolk — while stirring with a paintbrush. A little water is added to make the paint thinner and to facilitate application. Test the mix by painting it on a piece of clean window glass: if made correctly, it should peel off in a long continuous film. Egg white paints can be made similarly using the whites instead of the yolks.

While egg paints may not be practical for painting walls, they can be used for creative detail painting or as a glaze.

BEER GLAZES. Long ago, we're told, some resourceful painters stumbled upon a way to put left-over beer to good use: producing a pigmented glaze. When pigmented, beer forms a slow-drying glaze. It is typically painted over lighter colored paints to produce a wood grain pattern, achieved by dabbing and texturizing the glaze with paint brushes or sponges. When it dries, it forms a varnish- or shellac-like finish.

7-12: Common things in our household and natural environmental can often be used to produce the most beautiful finishes. Something as simple as the common chicken egg can be used to make paints.

To produce a beer glaze, pour the beer into a bowl, let it sit until it loses its carbonation, then and add enough pigment to color the glaze. Don't add too much, however: you don't want the glaze to become opaque. Mix the solution without producing bubbles, then apply the mix.

OIL PAINTS AND FINISHES. Another paint you may want to try making is produced from boiled natural linseed oil and pigments. To make it, pour linseed oil into a bowl. Swirl the oil around so it coats the sides, then add pigment, a little at a time, stirring continually until the pigment dissolves in the oil. Continue to add pigment until the paint thickens to the consistency of cake batter. Then add more linseed oil to produce a more liquid paint which can then be applied with a paintbrush. This simple oil glaze and other healthy natural oil-based glazes can produce uniquely creative finishes.

Other natural oil and wax finishes can produce deep, beautiful, and lasting

BEWARE!

Not all 100 percent linseed oils (boiled and raw) are free of chemical additives. Purchase the healthier versions of linseed oil for wall finishes.

surfaces. Natural penetrating oils, such as AFM Natural Penetrating Oil, Bioshield Penetrating Oil, and Bio Hard oil, can all produce durable finishes. OS Hard Wax oil natural wax finishes applied over a penetrating oil surface can be buffed to create a beautiful surface.

Paints can also be made from pigment using a number of other binding agents, such as starch, gelatin, and gum arabic. Recipes can be found in books listed in the Resource Guide.

Natural paints provide a means of changing the color of an earthen finish or refreshing a surface. They also provide additional protection while allowing walls to breathe. They're easy on our health and won't release harmful chemicals into the air we breathe. On top of this, they're all pretty easy to make! What more could you want?

7-10: You'd be amazed at what can be used to make a finish or paint. Even something as simple as stale beer mixed with pigments can make a beautiful glaze!

Quick Recap

Obviously, there are many ways to finish an earthen plaster wall. Alises are one of the most popular, but litema, decorative clay finish coats, and natural paints are also relatively simple and effective ways of adding color and texture to a wall. These finish coats can be colored by adding a variety of natural earth pigments which get their color from the minerals they contain. Whatever pigment you choose, be sure to inquire about its toxicity. Experimenting with pigments will help you achieve the desired color.

Alises, thin earthen natural finishes, provide protection and beauty to interior and exterior walls. Cooked flour paste is often added to improve the durability of the finish. Easy to make and apply, alises are inexpensive and breathable, that is, permeable to water vapor. They're often applied by paint brushes or sheep skin. Two coats are common and the final coat of alis can be buffed or rubbed with sponges or yogurt container lids to produce a smooth finish.

Litema is a wall finish made from fresh cow dung and brightly colored clays. Decorative clay finish coats are similar to alises, but thicker and more pliable and are applied by trowel. Like alises, they are typically applied in two coats and the final coat can buffed or polished to create a smoother finish.

Silica paints are used on lime and earthen plaster walls, either interior and exterior, to provide protection and color. Casein paints are made from dried milk protein which can be applied with or without pigments. Natural paints can also be made from egg yolks, egg whites, beer, and natural oils mixed with pigments.

In the next chapter, you will discover yet another finish, lime washes.

CHAPTER 8

Lime Plasters For Interior and Exterior Applications

L IKE EARTHEN PLASTERS, lime plaster has a long history of use. Historically, the first recorded use of lime as an exterior plaster is in Turkey around 4,000 BC — about 6,000 years ago. The Romans also used lime to make plasters as well as mortars (see Figure 8-1). In the cold, wet northern regions of Scotland and Wales, lime plasters were traditionally applied to stone buildings to protect the mortar from harsh weather and fierce rains. They often used a practice called lime "harling" — a process in which lime plaster is thrown onto a wall. This provided a "sacrificial coating" that prevented water from cracking and damaging mortar joints in stone walls.

Although it is still in use extensively throughout the world, in North America lime plaster has fallen into relative obscurity, largely due to the extensive use of cement stucco. Lime, however, is making a comeback, gaining wider use among natural builders who are looking for a sturdy, weatherproof means of protecting their natural homes from the elements.

This chapter explores the world of lime plaster — its uses, how it is made and applied, and its advantages and disadvantages — so you can make

8-1: Lime plaster used on this structure in Sesimbra, Portugal has protected the building for many hundreds of years.

an informed decision about using this marvelous but somewhat difficult material. Although we've tried to cover the subject comprehensively, we suggest you read and study further. Unlike earthen and gypsum plaster, there are a lot of books on lime plasters, and we also highly recommend that you take a workshop; we've listed books and workshops in the Resource Guide at the end of this book. Talk to experts for helpful hints, but keep in mind that expert opinions often differ — sometimes rather markedly.

Study everything you can, experiment, and then choose a system of lime plastering that works best for your project.

Understanding Lime Plaster

Lime plaster is a mixture of lime and sand. Although it is considered a type of natural plaster, lime as such does not exist in nature: it is manufactured from an abundant natural material, notably limestone, in a rather energy-intensive process.

As you may know, limestone is a sedimentary rock, deposited during several geologic periods by a variety of mechanisms. Most limestones are organic in origin, meaning they were derived from living organisms, more specifically small marine plants and animals. When these organisms died, they were deposited on the ocean floor in great number, forming thick deposits. Over time, the organic components of the organisms decayed, leaving behind a calcium-rich ooze, known as *calcite* or *calcium carbonate*.

As calcium carbonate accumulated on the bottom of oceans, magnesium carbonate precipitated out of the water, adding to deposits of calcium carbonate. Different amounts of magnesium carbonate were deposited in different areas, accounting for the three basic types of limestone, each with a different concentration of magnesium carbonate (see sidebar).

Over time, the calcareous ooze — containing varying levels of magnesium carbonate —consolidated into limestone, leaving little, if any, evidence of its organic origin. These deposits were buried by other sedimentary materials. Some areas of seabed became land during the long geologic history of the Earth. These land-based deposits are now mined to acquire limestone for building and a host of other purposes.

Limestone deposits are widely distributed throughout the globe, and thus widely available to many cultures, which explains the extensive use of lime worldwide throughout human history. Limestone is extracted from open pit mines, called *quarries*, then crushed and heated in kilns at high temperatures in excess of 2,000°F — one reason why lime is more environmentally damaging than earthen plasters (Figure 8-1). When limestone is kilned, the heat drives off carbon dioxide and water, converting calcium carbonate and magnesium carbonate into calcium oxide and magnesium oxide — commonly known as *quicklime*. (Its name comes from the fact that it reacts very quickly when added to water.

A NOTE ON TERMINOLOGY

Because lime plasters are made from lime and sand, they are often called *lime-sand plaster*. We use the terms lime plaster and lime-sand plaster interchangeably.

TYPES OF LIMESTONE

Geologists recognize three types of limestone: (1) high-calcium/low-magnesium limestone with 0 - 5% magnesium carbonate; (2) medium-level calcium/medium-level magnesium limestones, called *magnesian limestones*, which contain 5 - 35% magnesium carbonate; and (3) the dolomitic limestones, or simply dolomites, with high levels of magnesium, about 35 - 40%. High-calcium/low magnesium limestone is considered by many to be the best for making lime plaster.

Formation of LIMESTONE

Fig 8-1(a) (b): (a) Lime plaster is made from limestone. Crushed and heated in a kiln, limestone is converted into quicklime. It is hydrated to produce lime putty, which is then mixed with sand to produce lime plaster. (b) Insert illustrates the Lime Cycle.

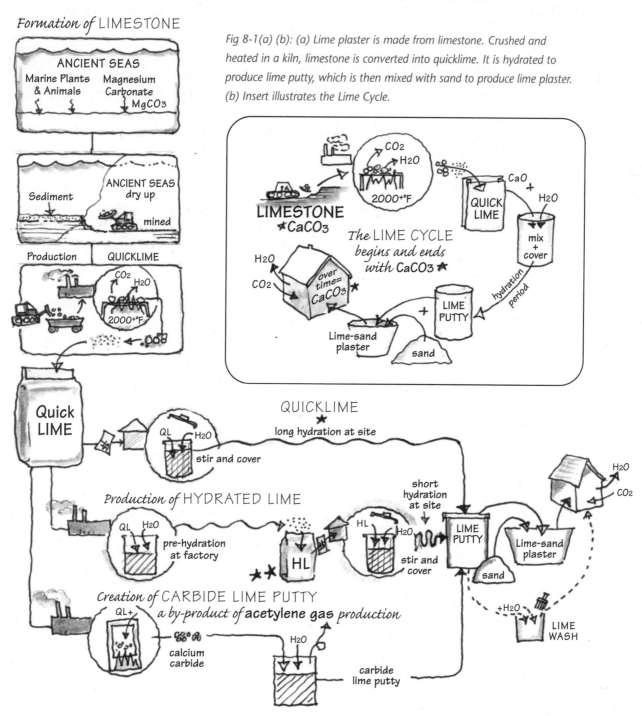

A LIME LEXICON

Limestone — the raw material from which quicklime is produced; calcium carbonate and varying amounts of magnesium carbonate.

Quicklime — made by heating crushed limestone. Dry powder used to make lime putty; calcium oxide and varying amounts of magnesium oxide.

Hydrated lime — partially hydrated quicklime; dry powder used to make lime putty; calcium hydroxide and varying amounts of magnesium hydroxide.

Lime putty — made from quicklime and hydrated lime; used to make lime plaster; slaked lime consisting of calcium hydroxide and varying amounts of magnesium hydroxide.

To make lime plaster, quicklime (available in powder) must first be hydrated — that is, soaked in water. In this process, also known as *slaking,* calcium oxide and magnesium oxide react with water to form calcium hydroxide and magnesium hydroxide, respectively. Huge amounts of heat are given off by the chemical reaction. The result is a gooey substance known as *lime putty* or *lime paste* — the primary ingredient of lime plaster — which has the consistency of cream cheese.

Lime putty cannot be used immediately: it must age (fully slake) before being mixed with sand to produce lime plaster. Although quicklime needs to stand for only two to three months before use, two years or more are preferable. In earlier times, when time seemed to fly by less quickly and people were less hurried, a five-year slaking period was deemed appropriate. In cultures in which lime plaster is still commonly used, long slaking periods are common. The longer the slaking, the better the putty: long-term slaking helps to fully hydrate the material, removing any lumps of calcium and magnesium oxide and making the lime putty smoother and easier to apply. More important, it also reduces potential damage after the plaster is applied, because unhydrated materials — calcium oxide and magnesium oxide — in the lime plaster will hydrate in the wall. When they do, they can explode, creating tiny pits in the wall surface.

After slaking, lime putty is mixed with sand to create lime plaster, or more precisely lime-sand plaster, which is then applied to walls. At this point, the calcium hydroxide in the mix begins to react chemically with carbon dioxide in the atmosphere. As a result, it is slowly converted back to calcium carbonate, forming limestone — a durable finish that could last hundreds, even thousands of years.

Lime plaster is applied to all sorts of natural buildings, including straw bale, cob, and straw-clay, either directly to walls or over a base coat of earthen plaster. But why use lime plaster?

Why Use Lime Plaster?

Lime plaster is used on interior and exterior walls, but its benefits are most evident in exterior wall applications where it offers superb protection against the weather, especially rain. Many cob homes in southern England are protected by lime plaster (see

8-3: Lime works well for interior and exterior walls, providing a durable protective coat that prevents mold in areas of high humidity.

Figure 1-8). Like earthen plaster, lime plaster is beautiful to behold. It can be pigmented to provide color or painted with a pigmented solution of lime called *limewash*, which provides color and additional surface protection.

Lime plaster is also vapor permeable or "breathable," a subject we discussed at length in Chapters 2 and 4. As Paul Lacinski and Michel Bergeron point out in *Serious Straw Bale*, "Vapor permeability serves as a safety valve for bale walls." By permitting water vapor to escape, straw bale or earthen walls can be protected for decades.

Not only does lime plaster allow a wall to release water vapor absorbed from various sources, it does not wick water as cement stucco does. In other words, it won't draw water into the interior of an earthen or straw bale wall.

Although there's a good deal of historical evidence showing that lime plaster protects earthen walls, there isn't much experience to draw on to assess its long-term performance in straw bale structures. The reason? Straw bale building is relatively new; newer yet is the use of lime in such structures. Based on its excellent performance in earthen structures, however, lime plaster should perform admirably over the long term, especially when applied over an earthen base coat, providing superior physical protection of walls against rain and wind while permitting walls to breathe — that is, release water vapor.

Another advantage is its durability: lime plaster is extremely strong stuff that has withstood hundreds of years of weather. Moreover, because it is pliant (somewhat flexible), lime plaster is not likely to crack very much as a building shifts or as walls expand and contract in response to natural temperature fluctuations or as moisture levels rise and fall. (In fact, lime is added to cement stucco and cement mortars to make them more resistant to freezing and less likely to crack.) In addition, lime plaster and earthen materials expand and contract similarly. Consequently, when lime plaster is applied on an earthen wall or an earthen plaster, it is less likely to crack or peel off a wall.

Lime plaster shrinks very little when it dries, which reduces potential cracking. In addition, because lime does not absorb moisture like cement stucco, it is less likely to damage underlying earthen materials. You will recall that cement stucco wicks moisture into walls, which contributes to the deterioration of earthen materials, and has a tendency to separate from the wall. Freezing and thawing of cement stucco also cause the stucco to deteriorate and crumble. Lime plaster, in contrast, is highly compatible with earthen plasters or earthen wall materials: they work in conjunction to produce a solid, protective finish.

LIME PLASTER VS. CEMENT STUCCO

Because lime plaster is durable and vapor permeable (breathable), lime is ideal for interior and exterior applications. Cement stucco, on the other hand, prevents water vapor from exiting, causing internal damage. Lime plaster has a lower embodied energy and tends to crack much less than cement stucco, and does not wick moisture as cement stucco does.

8-3: Here a worker applies lime plaster by flinging it onto the walls, a technique called "harling." Lime plaster goes onto walls easily, and needs to set slowly.

The net effect of these factors is that lime plaster tends to crack less than cement stucco. In addition, the fine hairline cracks that do form in a lime-sand plaster — for example, as a result of structural stress or from heating and cooling of the wall — are much smaller than in a cement stucco.

Lime plaster also has a remarkable capacity to self-heal — that is to say, hairline cracks seal themselves, as calcium hydroxide in the lime plaster reacts with carbon dioxide in the atmosphere and is converted to calcium carbonate — a process that naturally fills tiny cracks. These and slightly larger cracks can be repaired by applying a limewash, a milky solution of lime in water, discussed later in the chapter. Very large cracks need to be repaired with lime plaster.

Lime plaster is fairly easy to apply. Lime adheres well to straw and earthen wall surfaces and goes on without stucco netting. Being plastic (pliable) and adhesive, it trowels easily onto walls (Figure 8-3). Lime plaster is slow to set up. This allows the applicator lots of time to work. There's no need to rush. As a general rule, five to seven days are required to set the plaster "green hard" — that is, hard enough so that a thumbnail can be pressed into the plaster, but a knuckle can't. Lime plaster takes years to fully set or cure — that is, to convert entirely to calcium carbonate.

Yet another advantage of lime plaster is that its alkaline environment inhibits the growth of mold and mildew. These troublesome organisms can discolor walls and may cause health problems for the occupants of a home.

One final advantage of lime plaster is that it feels warm — much warmer than cement stucco in the winter. That's because lime has a lower thermal conductivity than cement.

LIME PLASTER MAINTENANCE

Periodic applications of limewash to lime plaster buildings have been used by many cultures to protect interior walls against mold and mildew, add color, refresh walls, and seal hairline cracks. Periodic applications to exterior walls seals hairline cracks and improves their durability and weather resistance.

Some Disadvantages of Lime Plaster

Although there are many reasons why you should use lime plaster, lime does have its downsides. For example, lime is a fairly caustic material, which can burn the eyes and skin, especially during preparation — for example, when quicklime is mixed with water to produce lime putty. During this heat-generating (exergonic) reaction, the boiling solution can splash lime onto bare hands, arms, faces, and into an applicator's unprotected eyes, causing burns — so be careful!

Although lime is less energy intensive than cement stucco, it is considerably more energy intensive than locally sourced earthen plaster. If you're concerned about environmental impact and ease of application, use an earthen plaster for exterior and interior finishes, if possible, or apply an earthen plaster base coat with a lime finish coat.

Lime is a finicky material, too. As noted previously, after it is applied in a lime plaster, calcium hydroxide carbonates — that is, it converts from calcium hydroxide to calcium carbonate after reacting with carbon dioxide. Carbonation occurs optimally under warm, slightly damp conditions; it slows down considerably if the weather is either too wet or too hot and dry.

Lime plaster requires some babying. After it has been applied, it must be wetted once a day for 7 to 14 days in cool, moist climates to ensure a good set. In hot, dry or windy areas, the walls may need to be misted two or three times per day to prevent them from drying out too rapidly. (If they do, carbonation slows down.) Tarps may need to be draped over the walls to keep the walls moist in hot, dry or windy areas. Burlap can be also be used to drape walls if it is wetted periodically (see Figure 8-5).

Although lime plaster's slow set time makes for easier application, it may pose some significant challenges, especially in colder climates with a narrow building window — that is, a limited time to build a home. In such instances, construction typically begins in the spring or early summer. Walls are built during the summer, and exterior finish work usually commences in the fall — when freezing temperatures are just around the corner. Lime plaster, like earthen plaster, can not be exposed to freezing temperatures until it has cured at least five to seven days. If lime plaster freezes before it sets up, it is likely to suffer permanent damage: the plaster will flake off the wall. Further complicating matters, lime plaster will not set at temperatures below 45°F (8°C).

Whereas heat and dryness slow down carbonation, cold weather stops the process altogether. If possible, schedule lime plastering to avoid cold weather; if not, tarps may be required to protect a wall at night from cold weather. Supplemental heat may be needed as well, though heat can cause uneven or too rapid drying.

If cold temperatures are approaching, you have several options. You can wait until the following summer to lime plaster the exterior walls — which is a good strategy if your walls are made from earthen materials or are earth-plastered, as in a straw bale home. Earthen walls can withstand winter weather, and an earthen base coat over an

> ### SAFETY FIRST!
> When mixing and applying lime plaster, wear full protective clothing, including gloves and goggles, so there's virtually no exposed skin. A good hat is also useful. Be careful not to scratch your skin with a lime-covered glove. Wear a certified dust mask when mixing powdered lime products, too. Lime is harmful to the lungs and mucous membranes of the respiratory system!

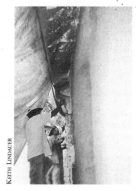

8-5: Rapid drying of lime plaster can destroy it. In dry climates, lime plaster must be misted one to three times per day for up to two weeks. Tarps or burlap may need to be draped over walls as well to prevent rapid drying. As shown here, workers may even need to apply lime plaster under the cover of burlap draped over the walls.

KEITH LINDAUER

earthen or straw bale wall will protect the walls from moisture and wind. When spring comes and cold weather becomes a distant memory, you can apply lime plaster.

In rainy climates, another option is to protect walls from weather, especially driving rains, with tarps or plastic until good weather arrives. Be sure to drape tarps loosely on the walls, away from the surface so internal moisture can escape, as trapped moisture will promote mold growth.

Yet another option is to speed up the set time by using additives. Many materials can be used to speed up the set time of lime plasters, some more environmentally benign than others. Pulverized brick, fired clay tile, and clay pots have been used in the past; these are first pounded or milled, then added to the plaster. Pulverized brick is sometimes available in local pottery supply outlets. Some lime plasterers also add fly ash, a waste material from coal-fired power plants, which is an effective additive, but dangerous stuff to work with as it contains many toxic heavy metals. Fly ash may pose a health threat to the occupants of a home, as well, and we don't recommend it.

Another additive is hydraulic lime, which contains clay that speeds up the set time. Cement may also be added to accelerate the set time. According to straw-bale builder Michel Bergeron from Canada, if the amount of cement in a lime plaster is held to under 20% of the volume of the lime in the mix, the set time can be increased to under 24 hours without compromising its permeability or strength. However, experienced lime plasterers from Europe shun this additive, feeling that it will weaken a lime-sand plaster. We agree with them.

Additives speed up the set by reacting with the lime to form crystals similar to calcium carbonate, though they form much faster than the calcium carbonate crystals. This makes the plaster harder sooner. Although additives may speed up the set time, they may reduce the vapor permeability of a lime plaster and weaken your finish. For straw bale and earthen buildings, this could be a dangerous tradeoff. (More on additives later.)

Lime, like earthen plasters, requires some maintenance. Although lime plaster does not crack much and is self-healing, it will need a fresh coat of limewash every two to ten years, depending on the severity of the weather. Some patching may be required as well. Limewashing and occasional patchwork take time, but much less than is required to maintain the exterior finishes of many conventional homes. In fact, it will take far more time to scrape peeling or chipped paint off wood siding and then repaint it, than to maintain a lime-plastered wall. Besides, all shelter requires some maintenance.

Finally, in North America it can be difficult to find trained applicators and materials. Call a local plaster company and ask about lime plastering your house or go to a local building supply center and ask for quicklime (calcium oxide) and lime putty (calcium hydroxide), and you'll quickly understand what we mean. Don't despair, though: we've listed sources of materials in the Resource Guide.

Making Lime Putty

Lime plaster is made from sand and lime putty. If you are lucky to live in a country where lime plaster is still commonly used, such as Great Britain, slaked lime putty can be purchased from construction companies or other suppliers, and then mixed with sand to make lime plaster. In North America, however, slaked lime putty is rather difficult to find. (See the Resource Guide for suppliers.) You can also make your own lime putty from either quicklime or hydrated lime, or use carbide lime sludge.

> ### LIME PUTTY IN THE UNITED STATES
>
> Lime putty can be purchased fairly readily in Europe. In North America, however, it is not widely available; that is, it is not generally available at conventional building supply outlets. North Americans can mail order lime putty from sources listed in the Resource Guide.

Making Lime Putty from Quicklime

Quicklime is a dry powder consisting of calcium oxide and variable amounts of magnesium oxide (depending on the limestone source). Although it is rather difficult to obtain in North America, many plasterers believe slaked quicklime produces the best lime plaster.

Slaking quicklime is a relatively simple process, but as the authors of *The National Lime Association Handbook* point out, "Like other apparently simple operations, it requires skill and experience." Both can be acquired with a little practice. However, an inexperienced person can ruin a batch of lime putty very easily — either "drowning" or "burning" it through carelessness or inattention. Quicklime can be dangerous, too: you don't want to inhale the powder or get it in your eyes or on your skin. Slaking quicklime — that is, hydrating it — while not life-threatening, can be hazardous as well.

Lime putty can be made from all types of quicklime, ranging from calcium-rich to magnesium-rich products. Whatever the composition of your quicklime, be sure to educate yourself about the process: read other publications on lime, and consult with an expert or two before you try making lime putty. Ideally, it would be best if you could work with an experienced lime plasterer who has made his or her own lime putty from quicklime before you try to manufacture your own.

Before you begin, determine the ratio of calcium oxide to magnesium oxide in your quicklime. As you'll soon see, this will make a huge difference in how you mix the material. Also, be sure to wear full protective clothing: essentially, cover all exposed areas of your body.

8-6: To make lime putty, be sure to use clean water and a clean vessel — a heavy-gauge hard plastic or metal container. This process gives off a considerable amount of heat. Most plastic garbage containers just can't take the heat!

Making Lime Putty from Calcium-rich Quicklime

Here's how lime putty is made from a calcium oxide-rich quicklime. (Separate instructions for a high-magnesium oxide quicklime will be presented shortly.)

To make lime putty, *clean* water is first added to a *clean* vessel — a bucket, wheelbarrow, or some similar container (Figure 8-6). Be sure to use a heavy-gauge hard plastic or metal container: most plastic garbage containers can't take the heat! Always start with at least three parts water to one part quicklime. When you are learning, it is best to start with a small batch.

After pouring the water into the container, slowly add the quicklime, which quickly sinks to the bottom. Continue to add quicklime slowly until the mix bubbles gently at a temperature of approximately 210° to 250° F. Be sure to stir constantly so all of the quicklime hydrates and all quicklime is completely under water. If necessary, add more water to cool it down. Bear in mind, however, that adding too much water will increase the amount of time required to obtain a paste of proper consistency.

Be careful: quicklime reacts quickly and can produce enormous amounts of heat. By adding material slowly, you can keep the temperature in the safe range.

Once the mix has stopped boiling, stir it with a clean hoe to ensure that all of the particles of quicklime come in contact with water — otherwise, they won't react! Now let the stuff sit for complete hydration. Remember, although you are dealing with a high calcium quicklime, it will have some magnesium in it, and the higher the magnesium content, the longer the hydration period. For optimal results, be sure to stir it once a day for the first three or four days and be sure to store the material in a closed container with a couple of inches of water over the top of the lime to keep air from reaching the putty.

Check the putty regularly and add water as needed. If the putty comes in contact with air, it will begin to carbonate — that is, to react with carbon dioxide. Carbonation in the bucket decreases the ability of the plaster to bind to the wall.

Making Lime Putty from a High-Magnesium Quicklime

Manufacturing your own lime paste from a high-magnesium quicklime is also pretty simple. However, rather than adding lime to water, water should be added to the powder. In addition, to slake

---

CAUTION!

Quicklime goes off quickly — that is, it reacts quickly and produces an enormous amount of heat. Be sure to keep containers of lime in safe areas, away from areas frequented by kids and animals. Cover containers securely to prevent curious fingers from getting into your putty.

a high-magnesium quicklime, you'll need only about half to two-thirds as much water as needed to slake a high-calcium quicklime. Spread the lime evenly along the bottom of the mixing vessel. Add water slowly to the powder at first, until slaking begins (that is, bubbles start emerging), at which time you can add more quickly.

Once all of the water is added — that is, the powder is covered by a one- or two-inch layer of water — stir the mix thoroughly so that every particle of lime comes in contact with water. This is where experience comes into play. Adding too much water to a magnesium-rich quicklime can drown the mix, making it ineffective: drowned lime will not harden, nor will the putty be smooth, viscous, and plastic. You can tell that you've messed up if the putty will not harden or the paste is watery.

Too little water can cause the putty to "burn" — this means that tiny particles of unslaked quicklime will end up in the plaster, preventing it from slaking fully. The particles fail to hydrate until the plaster is applied to a wall, and then explode, blowing tiny pits in the surface.

8-7: When storing lime putty for any length of time, be sure that there's a two-inch (5 cm) layer of water on top. This prevents carbon dioxide from reaching the lime (calcium hydroxide) and converting it back into limestone (calcium carbonate).

Screening and Storing Lime Putty

After lime putty made from quicklime cools, usually in a day or two for larger batches, it is screened to remove lumps or impurities. It is then stored in buckets or barrels with a one- to two-inch layer of water on top to prevent the material from coming in contact with air and starting to carbonate. If it does, it will become unusable. Check the mix frequently and add water, if necessary, to maintain the protective water layer (Figure 8-7).

One final note on making lime putty: Quicklime swells as it is slaked — by as much as 30% of the dry volume of the powder. Be sure your container is large enough to hold the expanded quantity of lime putty.

Keep 2" of WATER →

over the LIME PUTTY →

Making Lime Putty from Hydrated Lime

Lime putty can also be made from powdered hydrated lime. Purchased by the bag, hydrated lime is pre-slaked in the factory using a set amount of water, then dried using heat released from the slaking process. Hydrated lime products consist of calcium hydroxide or a mix of calcium hydroxide and magnesium hydroxide. These products are readily available and quick and easy to use. (Most building supply outlets sell hydrated lime.) Some plasterers contend that lime plasters made from hydrated lime do not perform as well as plasters made from slaked quicklime.

HYDRATED LIME

Hydrated lime is a preslaked lime powder available in bags. It contains calcium hydroxide or a mix of calcium hydroxide and magnesium hydroxide. Because it is preslaked, it takes much less time to make lime putty than from quicklime. Type S hydrated lime products are now used in over 90% of the construction projects utilizing lime. Be certain not to use agricultural lime for plasters. It is the wrong material!

Two types of powdered hydrated lime are available: Type N or Normal Hydrated Lime and Type S or Special Hydrated Lime. Type N lime is only partially hydrated and is generally a high-calcium product. There are some exceptions, however. GenLime in Ohio, for instance, produces a Type N hydrated lime that is derived from dolomite, a type of sandstone with a high magnesium content. Because high-calcium Type N is only partially hydrated, it requires additives or longer soak times for slaking (than Type S). That said, some Type N dolomite hydrates (magnesium-rich) have excellent working characteristics, even though they require 24 hours or more to soak.

Because of these factors, many plasterers use Type S hydrated lime products, containing a mix of calcium and magnesium hydroxide. Type S hydrated lime is more fully hydrated and therefore requires less time for slaking — usually only a period of 20 minutes or so. Type S hydrated lime holds water better than Type N limes and is more plastic. Good water retention provides good workability over absorptive surfaces.

In sum, hydrated lime is a powdered hydrated or preslaked product made from quicklime. Like the limestones they're produced from, they contain either pure calcium hydroxide or a mix of calcium hydroxide and magnesium hydroxide and can be used to make your own lime putty. All in all, Type S hydrated lime is generally considered a superior product, and is more widely available, so if you want a faster-setting product with excellent workability, use Type S. Type N dolomite hydrate works well, too, but it sets slowly. Whatever you do, be sure not to use agricultural lime! (It's commonly made from ground limestone and therefore consists primarily of calcium carbonate.)

MIX BY SIFTING POWDER INTO WATER. When making a lime putty from either type of hydrated lime, sift or slowly shake the powder into the water. Be sure someone stirs as you add the lime. As in any other process involving lime, you'll want to be careful, and wear full protective clothing, gloves, eye protection, and a certified dust mask.

Because lime putty production from Type N and Type S hydrated lime — as well as quicklime — can be somewhat dangerous work, we suggest that you consult *The National Lime Association Handbook* for further instructions or, if possible, work with an expert in your area the first time or two.

SLAKING HYDRATED LIME. After the hydrated lime has been added to the water and thoroughly stirred, it must be slaked for a while. Although Type S products can be applied after a 20-minute hydration period, to be on the safe side you should let the mix sit for

Above left: Community gathers annually for the spring maintenance of the historic adobe church of St. Francis, in New Mexico. Workers repair earthen plaster and apply alis protective coatings.

Above right: Every spring Djenne's Great Mosque is replastered. This spectacular event is a contest between the two halves of the town, whose teams work on the north and south sides on different days.

Lower left: The enduring beauty of natural plaster is evident in this building in Tunisia, where it has protected and enhanced the stone walls for nearly 1,000 years.

Right: Creative natural earthen and colored clay plaster finishes on a cob house.

Lower left: Playful sun sculpture of cob finished with natural earthen plasters.

Lower right: Litema plaster over an earthen base plaster provides rich and beautiful colors on this garden wall of cob and straw bale.

Top left: Earthen plasters provide beauty and protection to this straw bale home. Note the eave lattice detail which protects the plaster and the upper sections of the straw bale walls.

Top right: Earth plaster over cob and cordwood..

Lower left: Hand application of infill earthen plaster to the exterior of a straw bale house.

Lower right: Earthen plaster over cob and cordwood. Straw bale vaulted roof is finished with earthen plaster. Natural casein paint applied for color.

CEDAR ROSE GUELBERTH

CEDAR ROSE GUELBERTH

CEDAR ROSE GUELBERTH

CEDAR ROSE GUELBERTH

*Upper left & lower right: Layers of thinned paint, called "lesur",
applied over a gypsum plaster finish provide a depth of translucent
color to the straw bale walls of these Waldorf School classrooms.*

*Lower left and upper right: Alis finish with mica flakes
sparkles in this beautifully finished adobe building.
The alis is applied over a gypsum plaster base.*

Top left: A bright kitchen finished with pigmented lime plasters over cob.

Top right: A typical street scene in Portugal captures the common use of lime plaster finishes to adorn and protect the walls of these traditional homes, made from natural materials.

Lower left: This cottage features fresco painted lime plaster over cob walls.

Lower right: The gracefully hand sculpted cob walls of this living room are finished with lime plaster.

Top: Traditional lime plaster and lime wash finishes over earthen construction in Portugal.

Center left: This ridge is a good example of the rainbow of colors that can be found in nature. These natural clays and earth pigments are used in plasters and finishes.

Center right: Buckets of colorful naturally pigmented clay paints.

Lower left: Final color coat test patches over earthen plaster, on tire walls.

Lower right: Carole Crews buffing an alis coat over an earthen plastered adobe wall.

POVI KENDAL ATCHISON

KEITH LINDAUR

POVI KENDAL ATCHISON

CEDAR ROSE GUELBERTH

Top left: Pigmented natural clay plaster applied over straw-clay walls enhance the beauty of this traditional timberframe. Sculpted Niche and arch use cob and willow for sculpting base.

Top right: Natural paints brighten up the finish of this Earthship. These paints are applied over an earthen plaster coating the earthship walls.

Lower left: French ochre pigmented natural clay plaster over earthen plaster on straw bale walls.

Lower right: Natural earthen plaster over straw/clay construction provides beauty and protection to this office.

Upper left: Sienna and burnt umber pigmented clay color coat plasters, over earthen plaster base on tire walls.

Upper right: Newman red clay provides a warm tone to this natural clay plaster applied over an earthen plaster base, on a straw/clay wall.

Lower left: Sculpted cob window seat. Straw bale walls, with natural clay finish plaster over earthen plaster base.

Lower right: French ochre pigment in natural clay plaster over earthen plaster on straw bale walls

at least one full day to ensure complete slaking. (Be sure that there's a layer of water covering the lime at all times!)

Longer periods of slaking make the putty stickier and more plastic and thus easier to work with. If you can, let the putty sit for a few weeks: you'll end up with a better product. A few months would be even better; a couple of years would be ideal. Remember: for all handmade lime putties, the longer the slaking period, the better the product. Be sure to mix the putty once a day for the first three or four days.

Hasty slaking of hydrated lime (and quicklime) sometimes leaves small pieces of unslaked lime in the putty. As noted earlier, these hydrate if the putty is allowed to stand for long periods. If permitted to get into the finished work, these granules of unslaked lime will react with moisture, popping and blowing small chunks off the wall. The result is a pitted surface.

> ## IMPORTANCE OF SLAKING LIME
> The object in aging putty made from hydrated lime is to assure that the particles of hydrated lime may have full and ample time in which to absorb the maximum amount of water. This insures smoother putty of greater workability and sand-carrying capacity.
>
> *The National Lime Association Handbook*

Lime putty can be made in small to large batches, whatever suits your purposes. To make small batches, use a clean five-gallon bucket, a wheelbarrow, or a clean, shallow mortar boat. To make large batches, you may want to mix in a mortar mixer and slake in large clean buckets, clean heavy-duty metal or heavy-duty, hard plastic trash cans, or clean 55-gallon drums. (Be sure the container can withstand the heat.) Once the lime putty is slaked, it can be used to make plaster. Begin by draining off excess water. Scoop out some of the putty and transfer it to another clean container. Add sand and mix.

The use of powdered hydrated lime is the subject of debate. Some people dislike this product and recommend against it; they say it does not produce the best quality lime plaster. Others use it with great success, the key to which may be that they use fresh powdered hydrated lime — that is, the bags of powdered lime haven't been sitting around for long periods, undergoing carbonation. (As a reminder, carbonation is the formation of calcium carbonate, in this case, in powdered hydrated lime.) Carbonation decreases the bonding qualities of lime plaster, so be sure to purchase freshly manufactured hydrated lime to make lime putty. Remember: carbonation should occur on the wall, after the plaster has been applied, not in the bag or in the slaking bucket, as could occur if the layer of water over lime putty is allowed to evaporate.

Carbide Lime Slurry: A Substitute for Lime Putty

If you can't acquire lime putty, and don't want to make your own from quicklime or hydrated lime, one possible option is carbide lime sludge — a waste product from the production of acetylene gas which is used in welding.

Acetylene producers in the western United States begin by mixing calcium oxide (quicklime) with carbon (coke or anthracite coal) in an air-free furnace. (In the Eastern United States, acetylene producers use a dry process involving natural gas.) This produces calcium carbide — a dry, granular product that looks like gray gravel, also called *carbide lime* — which is mixed with water to produce acetylene gas and calcium hydroxide (hydrated lime) — lime putty. It is called *carbide lime sludge, carbide lime slurry,* or *generator slurry*: these are the names to use when calling manufacturers about acquiring some lime sludge.

CARBIDE LIME

Harry Francis, one of the nation's leading experts on lime plaster, notes that carbide lime putty resulting from the production of acetylene gas is a fairly pure calcium hydroxide with cementitious qualities that can be suitable for making plasters.

Carbide lime slurry is a fairly plastic (pliable) hydrated lime, generally stored in tanks or, less commonly, waste ponds, at manufacturing facilities. From here it is periodically removed and sold or disposed of in landfills. Carbide lime putty has the consistency of drywall compound or sour cream.

"Carbide lime putty possesses all of the same attributes as traditional lime putty," note Kaki Hunter and Doni Kiffmeyer, natural builders who specialize in earthbag construction and natural plasters and finishes. Hunter and Kiffmeyer assert that carbide lime is the closest thing to traditional pit lime (lime putty) available in the United States.

Carbide lime sludge is a well slaked putty, which is chemically consistent and has a high-calcium and low-magnesium content. Many lime plasterers believe a high-calcium, low-magnesium lime plaster is best for lime plaster.

Bear in mind, however, that carbide lime sludge is as caustic as lime putty and should be carefully handled using full protective gear, and stored in tightly sealed containers with a layer of water on top.

Carbide lime sludge has a slight grayish blue tint and "smells like a combination of ammonia and powdered garlic," says Hunter. Fortunately, the odors disappear as the material dries on a wall. Whether or not these smells are harmful to breathe, we can't say with certainty. We recommend caution, however.

Carbide lime sludge may also contain small specks of carbon., which are generally of no concern, except in finish coats. "All in all," remark Hunter and Kiffmeyer, "[carbide lime sludge] has proven to be a very clean, bright-white curing, pure source of lime putty, when properly harvested."

Carbide lime has been used continuously to make acetylene gas since the 1950s. In Moab, Utah where Hunter and Kiffmeyer live, there are two sources of sludge — small welding operations that make their own acetylene and produce carbide lime sludge! But they warn, "It is exceedingly rare to find small welding operations that produce their own acetylene in the United States. Most acetylene comes from large gas manufacturers such as Praxair or U.S. Welding," which may make it more difficult for ordinary citizens to acquire the material for personal use.

If you do locate a local source, be careful when mining waste lime sludge. Be sure to wear protective gear and tread carefully around lime pits: you don't want to fall in. Also, be sure to skim off the top layer, which is exposed to the air and thus partially carbonated. Acquire the putty from beneath the top inch or so, it is generally suitable for use.

Remember, although carbide lime is hydrated, you should probably let it age — that is, undergo further slaking. Three months is a good period, unless it has been in the pit for a long time.

Making Lime Plaster from Lime Putty

Lime plaster is made from lime putty and sand. As you have seen, lime putty may be derived from a variety of sources. It can, for instance, be purchased from suppliers or made from quicklime (powdered calcium oxide) or from hydrated lime (powdered calcium hydroxide). You may also be able to obtain some from carbide lime sludge pits.

Lime putty needs to be made long before plastering begins, so it is fully slaked when needed. Don't worry about shelf life, though: lime putty can be stored for long periods without damage, as long as it is covered by a one- to two-inch layer of water. Sealed buckets or barrels work well as storage containers. To maintain the putty's consistency, prevent it from freezing — in other words, prevent particles from separating from the water. If it does freeze, you'll have to remix it after it thaws.

Mixing Lime Plaster

Lime plaster is applied directly to walls or over an earthen plaster base coat, in one to three coats. One or two coats are applied as a finish coat over a scratch and brown coat of earthen plaster. Three coats of lime plaster are applied directly to a wall surface. We'll describe the three-coat system first.

In a three-coat system, the first two coats are generally made by mixing one part lime putty to two to three parts sand. Fiber may be added to increases its tensile strength. The finish coat has a slightly higher ratio of lime to sand, about 1.5 parts lime to two to three parts sand. Fiber is generally not added to the finish coat.

Sand adds strength to any plaster and reduces shrinkage. Sand particles in lime plaster should be fairly sharp, not smooth or rounded, to ensure maximum strength. Grain size should also vary within your mix to promote maximum strength. Very fine sand and sand of a uniform size will actually result in a weaker plaster — one that is more likely to crack.

Small, rounded sand particles also produce a weaker plaster that is more likely to crack or rub off. Beach sand containing salt will cause white blotches to form on the wall, as the salts migrate out of the plaster. (If you use beach sand, be sure to wash it well first.)

When making a lime plaster, mix water, lime putty, and sand together by hand (using a stirrer of some sort, not bare hands) or by machine. Most people find that a mortar mixer works much better than a cement mixer. In a cement mixer, lime plaster tends to stick to the drum, which must be periodically scraped clean to keep operations running smoothly. Both electric and gas-powered mortar mixers are available. For best results, first shovel half of the putty and half of the sand into the mixer. After the putty and sand have mixed, add the rest of the material.

After the components are well-mixed, reinforcing fiber may be added, although this is optional. Hair or chopped straw work well, but are generally added only to interior lime plasters and exterior base coats to increase their tensile strength. Don't add fiber to the finish coats — the lime plaster becomes too difficult to trowel. Hair and other fiber from the finish coat also absorb moisture, which could cause the plaster to crack in freezing weather. A wide assortment of fibers and animal hairs can be used. In Europe, horse hair is most commonly added to lime, but not hair from the mane or tail, which is too springy. Cow, goat, and pig hair also seem to work well. Human hair may suffice, so long as it is coarse.

No matter what type of hair you use, be sure to separate the strands — tease them apart — so the hair doesn't clump in the plaster. How much hair do you add? About one-half of the volume of lime putty — that is, if you're adding a bucket of lime putty, you'll need a half bucket of hair.

Because lime requires a lot of hair, some plasterers use straw for interior lime plasters and exterior base coats, as straw is easier to acquire in sufficient quantities. Use the same proportion as hair and be sure to chop the fibers finely into 1- to 3-inch (2.5 - 7.5 cm) pieces. More water may be required when straw is added because straw tends to absorb water and make the plaster dry. A good mix is usually achieved in about ten minutes.

Although some people worry that lime may eat away at straw, it appears that the lime actually causes the straw to calcify. Like petrified wood or fossil bones, the calcium fills the empty spaces inside the straw fiber and is converted to calcium carbonate, creating a very durable material.

When thoroughly mixed, a lime-sand plaster should stick to a trowel when it is turned upside down (Figure 8-6). If the lime plaster is too wet, it will likely develop shrinkage cracks, which weaken it.

8-8: You know your lime plaster mix is right if it sticks to a trowel when turned upside down. If it drips off, it is too wet and will very likely shrink and crack as it cures.

Applying Lime Plaster to Straw and Earthen Walls

Lime plaster can be applied directly to earthen and straw bale walls or over a base coat of earthen plaster. Our preference is for the latter.

By Hand or By Trowel.

Lime plaster, like earthen plaster, can be applied to walls in a variety of ways: by hand, trowel, or machine. Lime plaster can also be smeared on by hand, like an earthen plaster, but be sure to wear rubber gloves. Hand application creates a rough surface that provides a good key system for the next layer.

Some people like to throw the plaster on the wall by hand or by using a special "harling" trowel (Figure 8-9). Standing about three feet from the wall, a harler simply throws the material onto the wall one trowelful at a time to achieve good, even coverage.

Harling produces a rather rough surface. Because of this, some plasterers use a wooden trowel to knock down the high spots and even out the surface. Harling is difficult to master, but with practice and patience, even coverage is possible. Lime plaster can also be applied using a stucco pump.

Although lime plaster application is less hazardous than slaking quick lime, it does carry some risk. As noted earlier, lime is caustic, so your body should be well protected. Waterproof rubber gloves are recommended no matter how you apply it, and goggles are an absolute necessity. After a day's work, be sure to wash your hands thoroughly with soap

8-9: One technique for applying lime plaster is called "harling." To harl lime plaster, first stand about three feet away, then throw the mix onto the wall. Harling requires a special trowel, and considerable patience and practice, to achieve even coverage.

and water. When they're clean, rub in a little vinegar. The acetic acid in vinegar will neutralize lime, and virtually eliminate the dry, cracking, painful skin problems you may encounter.

Applying Lime Plaster over a Base Coat of Earthen Plaster.

As noted, our preference in straw bale and earthen-walled homes is to build up an even, durable earthen plaster, then cover it with a lime plaster finish coat. Follow the steps outlined in Chapter 6. Be sure to trim straw bale walls and apply an adhesion coat over wood, then apply a slip coat on the bales. Pack voids between the bales with dry straw and straw coated with clay slip before you apply the discovery coat. When that has dried, apply the infill coat; let it dry, and apply one to two lime finish coats. (Remember to wet the walls between coats of plaster and before applying lime plaster.)

This system allows you to apply the more user- and environmentally-friendly earthen plasters to build and shape a wall surface, while minimizing the use of the more difficult, more costly, and much higher embodied energy/higher environmental impact lime plaster. In other words, you can benefit from lime's durability, while minimizing its use. Lime bonds well to earthen materials, so there's no sacrifice in durability. Applying earthen base coats also reduces time, as earthen plaster can dry and cure faster between coats than lime.

One final note: do not overwork lime plaster — that is, don't smooth it out too much with your trowel, as this causes the sand to separate from the lime, significantly weakening the plaster.

Applying Lime Directly to Walls.

When applying lime directly to straw or earthen walls, be sure to wet the walls first. Straw bale walls should be trimmed and plumb, but there's no need to apply stucco netting. Netting can actually reduce the bonding of the plaster to the wall by creating air pockets at the interface of the wall and plaster, which may cause the plaster to break down.

Lime plaster, if applied well, keys nicely into straw, which is best accomplished by hand, by harling it onto a wall, or by using a stucco pump. If the bales are dense — and the fibers are tightly knit together — you may be able to apply the first coat of lime plaster by trowel, using a good sticky mix while being sure to key the material in well. If the bales are less dense, it's best to use a stucco gun.

Earthen walls generally require little preparation for lime plastering. Earthen plaster should be fully dried and wetted down prior to receiving a coat of lime plaster; use enough water to moisten the walls yet avoid water running down the walls. Allow the water to absorb into the wall for a minute or two, then begin applying the lime plaster. All subsequent coats of lime should also be wetted prior to application of additional lime.

Three-Coat System.

Lime plaster applied by itself is generally laid down in three coats, starting with the base coat or scratch coat. Typical thickness is $^3/_8$- to ½-inch. When using a trowel to apply a base coat, be sure to go back over the wall with a scratching tool. Scratch the wall about an hour after the first coat of lime plaster has been applied. Let it sit for a week or two before applying the second coat. When applying the lime by hand or harling it onto the wall, the surface should be rough enough to key the second coat in.

8-10: The scratching tool is used to rough up the first coat of lime plaster, appropriately called the scratch coat. This process helps ensure a tight bond with the next coat of plaster, the brown coat.

CHRIS MAGWOOD & PETER MACK

No matter what technique you use to apply plaster, you'll need to mist a new coat of lime plaster one to three times a day for one to two weeks, depending on how hot, dry, or windy the weather is. You can use clean tap water or lime water to mist the wall. Lime water comes from one of two sources: the "top water" from the lime putty storage containers, or by mixing one cup of lime putty with five gallons of water. Lime water speeds up the carbonation process. Be careful when you apply it not to splash the stuff on your skin or in your eyes.

Besides misting the wall, you may also need to drape a tarp over the wall to protect it from wind, dry air, and hot, baking sun. As noted earlier, you should avoid application of lime plaster when cold or freezing weather is imminent.

The second coat of lime plaster fills the voids, creating a smoother surface. You can't build the plaster too thickly or it will slump. Too thick a layer will also reduce contact with carbon dioxide in the atmosphere, which is vital for carbonation. If you're using a pure lime-sand plaster — that is, a plaster without additives to increase the set time — the second coat should be no thicker than $^5/_{16}$-inch. After the second coat is on, let it cure a bit, then retrowel it to remove cracks. Retroweling generally takes place after the wall has set green. After the wall is troweled, you will need to go over the surface with a sponge or sponge float, making circular motions. Burlap works well, too. This step helps to roughen the surface to ensure a better key-in with the finish layer.

Like the base coat, the second coat will require periodic misting with water or lime water one to three times a day over a period of a one to two weeks. The drier the conditions, the more misting is required.

If time is short, you may want to apply an earthen plaster base coat over the wall or over particularly irregular parts of the

Apply lime plaster in thin coats to better expose each layer to the atmosphere, which will facilitate hardening.

Matts Myhrman and S.O. MacDonald,
Build it With Bales

8-11: Worker troweling lime plaster on wall. Be sure to wear goggles or glasses, gloves, and cover up the rest of your body.

wall prior to lime plaster application. This creates a tight bond between the wall and the plaster and helps fill the dips and valleys more quickly. It also speeds up the process. You can apply a lime plaster once your earth plaster is dry — usually within a few days. This can cut weeks off a plaster schedule.

The final layer, containing 1.5 parts lime to 2 to 3 parts finer-grained sand, is the finish coat — a thin layer, typically no more than $1/8$-inch, which is applied by trowel, left alone to set, then retroweled to create a compact, durable water-resistant finish. Work the surface with a trowel, tracing tight circular motions while applying even pressure, but don't begin this process until after the finish coat has become firm. If the plaster has become too dry, you will need to brush, sprinkle, or mist some water on it beforehand. You will very likely have to wet the trowel also, to help it slide across the surface of the finish plaster.

If you want a rough texture, you can work the surface with a wooden float or a sponge float. This creates an open, grainy texture.

Lime plaster may be applied over flashing and wood. There are several ways this can be achieved. For example, wood can be coated with an adhesion coat made from flour paste, water, sand, and manure, as explained in Chapter 6. Lime plaster can then be applied directly to the dried adhesion coat.

Another option is to apply a layer of earthen plaster over wood, as described in Chapter 6. After it has dried, wet the surface, then apply a coat of lime plaster over it.

Additives to Lime Plaster

Lime plaster works really well on its own, especially when made from a well-slaked lime putty. However, lime plaster can be improved upon by adding various materials, a few of which were mentioned earlier in the chapter.

One natural additive worth considering is cactus juice, a material we discussed in connection with earthen plasters. Cactus juice reduces the set time and improves the quality of a lime plaster, making it more water repellent and increasing its durability. Casein powder and skim milk also make lime plasters more durable and water-resistant. Manure makes lime plaster more plastic, and thus easier to apply to a wall. It also makes it more weather resistant. Gypsum accelerates lime's setting time and makes lime plaster stronger early on, so that additional layers can be added more quickly. Gypsum also reduces shrinkage. Marble dust, quartz dust, and mica can be added to create a more durable finish and to enhance the beauty of the walls.

Numerous other additives have been used as well. Pulverized brick, fired clay tile, clay pots, cement, and fly ash additives, for example, all reduce the set time. We recommend that you stay away from the potentially toxic fly ash and high embodied energy cement, for reasons enumerated earlier.

Lime plaster can be sealed with potassium silicate (waterglass), natural oils, or beeswax to prevent stains and to improve water resistance. However, your best bet is to seal a wall naturally — that is, to allow the natural carbonation process to occur. Carbonation produces a well-sealed, but vapor-permeable or breathable wall. Limewashes (discussed shortly) enhance this process, and also allow for breathability. Whatever you do, never seal lime with a synthetic sealer. It will impair the breathability of a wall.

> ## MARBLE DUST
>
> Marble dust can be added to the final coat of lime plaster; this makes the plaster more workable and when dry creates a hard, reflective surface. However, you'll need to reduce the amount of sand you add to the final coat of lime in direct proportion to the added marble dust.

Coloring the Finish Coat

Lime plaster walls appear white when dry, but are often colored to produce a more stunning appearance (see color gallery). Coloring may be achieved by adding pigments to the finish coat mix and/or by applying a pigmented limewash, a fresco paint, or some other natural paint over the finish coat (see Chapter 7).

When pigmenting a lime plaster, applying a fresco paint, or limewashing walls, be sure that the pigments are compatible with lime. Lime is a base (that is, it has a high pH) and reacts with acidic pigments (low pH). In addition, some pigments react with salts or metals in the atmosphere. As a result, not all pigments have the stability to remain permanent over long periods.

As a general rule, all of the iron oxides, which include yellow, orange, and red ochers, and the blue cobalt pigments are stable, although cobalt blue is highly toxic. Some ultramarines and copper-based colors do not work with lime. (For more on pigments, see Chapter 7.)

Check out suppliers listed in the Resource Guide for more specific lists of pigments suitable for fresco painting and lime. You may also want to consult with a reliable pigment specialist for more information. Remember: all colors should be tested for stability before use. Six month's exposure to the atmosphere is generally required to determine if a pigment is stable.

You can also use synthetic or artificial pigments to colorize lime plaster. However, some artificial pigments are toxic and pose a health risk. If you want to use them, be sure to research them carefully. Eliminate those that could cause health problems. Remember, too, certain natural pigments may also be toxic; these should be avoided.

Some lime plasterers pigment their finish coats with tempera paints added to the lime putty. Add no more than three percent of the weight of the lime. Any higher and the pigments may interfere with the set and will surely weaken the lime plaster. The same holds for synthetic or natural pigments.

When making a pigmented lime plaster finish coat from lime putty, mix the dry pigments with sand, then add the sand and pigment to the lime putty and stir. Pigments can also be mixed with warm water, which is then stirred into the lime putty. Sand is added next.

Limewashes

Lime wash is a simple paint popular in the past and still admired for its matte finish and chalky texture. A limewash is made from water and lime putty, and applied to lime-plastered walls to provide protection and to seal cracks and porous surfaces. Without pigment, it dries a light white; when pigment is added, it provides protection and unrivaled beauty.

8-12: Limewash is typically applied by brush, as shown here.

Limewashes make lime plaster even more durable: they shed rain and thus protect the underlying layers of lime plaster from the elements. Even if you weren't going to pigment a lime-plastered wall surface, it's advisable to apply a limewash for added protection (Figure 8-12).

Limewashes are also a good idea if lime plaster is applied rough, either by hand or harling. Limewash essentially deposits lime particles in the uneven surface, "enough to prevent moisture from being trapped, but still allowing evaporation to take place easily," according to veteran straw bale builder, Bill Steen.

Limewashes are also used to color walls and produce a variety of soft, creative colors of extraordinary beauty. Pigmented lime washes can be applied in layers and may provide a deeper hue than a pigmented finish coat. They can even be applied over pigmented lime finish coats.

Limewashes can be applied to earthen and gypsum plasters, too, although you should experiment before applying over an earthen plaster to be sure it produces the look you want. Limewashes can make the earth-plastered wall look muddy unless well done. If you want a clean color or a brighter look, you should apply a lime finish coat first, then paint it with a pigmented limewash.

This watery natural paint serves a number of other important functions — besides tinting a wall and making the surface more durable and weather resistant. It may help to even out variations in the wall surface and mask cold joints — areas where one day's work stopped and the next began. When applied to a wall surface, calcium hydroxide in limewash converts to calcium carbonate, which fills hairline cracks in the finish coat and openings in porous surfaces, creating an extremely durable surface that remains breathable.

> ## BENEFITS OF LIMEWASHING
> - Seals cracks and porous surfaces
> - Creates a breathable surface
> - Adds color (when pigmented)
> - Refreshes and revitalizes surface of lime plaster

Making a Limewash

Limewash is made from well-slaked lime putty and water. Begin by adding water to the lime putty (about 1.5 parts of water to 2 parts lime putty) until it reaches a consistency of whole milk. (It will not be as white as milk, however: the mix will appear more translucent.) Next, run the mix through a sieve to remove any lumps or undissolved particles of lime. Pigments are mixed separately in water and then added to the lime-water mix, which is then stirred.

Knowing how much pigment to add can be tricky because limewash dries much lighter. To test the final color, apply a little on the wall and then let it dry. If it is too light, add a little more pigment. Be careful, however, too much pigment in a limewash may reduce its binding strength. And remember, to achieve a final color, you will very likely have to apply two or more coats of pigmented limewash.

Applying Limewash to Walls

When the limewash is ready to use, wet the wall thoroughly and evenly. If you apply limewash to a dry wall, the water will be sucked out immediately, impairing the bonding of the material.

Limewash is typically applied by brush. Limewash brushes come in many varieties: with natural or synthetic bristles, or a combination of the two; with short or long bristles. Experienced applicators generally find that nylon bristles hold up better than natural ones. However, a natural brush holds this very thin material much better than a synthetic brush, which means there will be fewer trips to the bucket and fewer drips on the way to the wall. It also means more even coverage.

Be patient when working with limewash: resist the temptation to work too quickly. Limewash is watery and much runnier than paint and tends to splash. When applying limewash, protect yourself by wearing rubber gloves, goggles, long-sleeved shirts, and long pants — covering all exposed skin.

Whether limewashing a new wall or refreshing an older lime-plastered wall, begin by removing dust and loose particles, then wet evenly with clean water. Remember that limewash is not as opaque or thick as normal house paint — in fact, it looks transparent when first applied. To obtain the desired effect, you'll need to apply several thin layers. Pigments in each coat can add amazing depth to colors. Each coat applied is almost invisibly thin, so that it will not flake off. Coat the surface evenly and work the wash in well. As it dries, limewash turns somewhat opaque. Three to six thin coats are typically applied. Let each coat dry before the next one is applied but be sure to wet down the wall before applying another coat.

Although you may be tempted to speed up the process by adding a thicker coat, don't do it. If applied too thickly, limewash will crack and flake off the wall. It may also crack and flake off if it dries too quickly — for example, when applied on walls in a hot, dry climate. If you're in such an area, you may need to protect the wall from hot sun and mist the wall several times a day for two or three days after each application of limewash.

For best results, wet the wall down the day before you are going to apply the limewash, then mist the wall just before you begin work. Successive coats should be applied several days apart. (Although a coat of limewash dries in a day or so, each coat will take several weeks to fully harden.)

Because pigmented limewashes lighten as they dry, run some test patches before you apply your wash to an entire wall. You can burnish each coat with a clean, damp brush after it has begun to set. A burnished limewash surface takes on a smooth silk-like texture.

On exterior walls, some applicators add a small amount of fine sand or marble dust to the first coat of limewash to create a more adhesive layer for successive coats. Some lime plasterers also add oil (such as linseed oil, castor oil, or hemp oil) to the final coat of limewash to increase water resistance (or apply it over lime plaster itself). These oils degrade over the first year, as the coat carbonates. So why use them?

Oils are used to provide immediate protection. However, because oils seal walls from the air, they reduce the rate of carbonation in the underlying lime plaster. As a result, oil-limewash should be applied at least six weeks after the lime plaster finish coat has been applied. We recommend extreme caution when using oil. If you can do without it, you're probably better off.

Fresco Painting

In Europe, Asia, Mexico, South America, artists once painted intricate murals (often on church walls and ceilings) using water colors — pigments dissolved in water and applied to lime plaster. Many of these paintings have survived for more than 3,000 years. The secret to their longevity lies in their application: pigment was dissolved in water then painted onto a wet lime plaster, which acted as the binder for the pigment.

8-13: This cob baking oven is finished with a lime-sand plaster and natural earthen pigments applied fresco style.

As the lime carbonated and the moisture evaporated, the pigment became locked in a fine crystalline layer that formed on the surface — which accounts for their durability. As Jocasta Innes, author of *Applied Artistry* and *Paint Magic* notes, "Painting in fresco was notoriously demanding: wet color dries many shades lighter ... so it takes a trained eye to determine the color values of a completed fresco, which might take weeks to dry."

Fresco painting was developed in Italy during the Renaissance, and is known by its Italian name, *buon fresco*, or true fresco. This challenging art form required a smooth, wet, white, well-compacted (troweled) base, prepared one section at a time. Three layers of plaster are required for true fresco. Another technique, known as *dry fresco*, involves the use of pigments in a binder such as egg, gum, or casein (milk protein), as described in Chapter 7. The mix is applied to dry lime plaster.

Today, fresco paints are used to color lime plaster walls. Lime-stable, water-soluble, and natural pigments are mixed with water, then painted over a wet lime plaster finish coat. The less water in the pigment, the more opaque the dried fresco finish will be.

How do you know if the final layer of plaster is sufficiently dry? About an hour after applying the plaster, run a wet paintbrush across the surface of plaster: if the water is absorbed within a minute's time, the wall is ready to paint.

As in the fresco paintings of an earlier time, the pigment becomes locked up in the surface layer as the lime carbonates and the water evaporates. Encapsulated in calcium carbonate, the pigment is no longer water soluble.

To enhance the beauty of the wall after fresco painting, the surface can be polished using a damp, clean paint brush. The result is stunning — vibrant and alive — as the white surface of the lime plaster illuminates the color from within, according to Bill Steen.

FRESCO SUCCESS

First layer of lime plaster: 3 parts coarse, sharp sand to 1 part lime putty

Second layer of lime plaster: 2 parts coarse sharp sand to 1 part lime putty

Third layer of lime plaster: 1 to 1.5 parts fine, sharp sand to 1 part lime putty

Lime-compatible pigments

Source: Annie Sloan, *Paint Alchemy*

Not all types of lime putty are suitable for fresco work: high magnesium or dolomitic lime putty, for instance, is not recommended due to its high magnesium content. The best lime putty is a well-slaked high calcium content material containing less than five percent magnesium. Lime plaster made from this material creates brilliant frescos with permanence, largely due to the high binding power of aged lime.

Quick Recap

Lime plaster is a mixture of sand and lime putty. Available through a limited number of commercial outlets, lime putty can also be made by mixing water and quicklime or hydrated lime, which is available in many building supply outlets.

Although it is finicky and much more dangerous than any other natural plaster, lime plaster is beautiful and is the most durable natural plaster around. It cures gradually over decades, producing a hard limestone-like finish. Because of this, lime plaster is often applied to exterior walls in areas that experience frequent driving rains.

Like other natural plasters, lime plaster is breathable and is therefore ideal for a wide variety of natural building systems. Be sure to wear protective eyewear and clothing, as well as gloves, when making lime putty or lime plaster and when applying lime plaster.

Lime plaster, while challenging, is well worth the effort. It creates a durable yet beautiful wall.

Now that you're ready to begin your adventures with lime, we strongly recommend that you consult other books and chat with experts on the subject. Lime plaster is a bit dangerous to work with and there's considerable room for error: use caution.

CHAPTER 9

Gypsum Plaster for Interior Walls

GYPSUM PLASTER has been used as an interior wall and ceiling finish for thousands of years throughout the world. You may be surprised to learn that gypsum plaster was used to finish the interior walls of the massive pyramids of Giza in Egypt built around 2,500 BC, 4,500 years ago.

In more recent history, gypsum plaster has been used to finish interior walls of many new homes built in both the United States and Canada, where it was applied by hand over wood lath. Slowly but surely, however, this type of gypsum plaster application has been replaced by drywall with gypsum texturing, a less labor-intensive method. Although drywall is made from gypsum, a wall built from drywall and texturized with gypsum is radically different from a wall with gypsum plaster over lath (Figure 9-1).

Today, drywall has gained in popularity and has virtually eliminated gypsum wall plaster. Nonetheless, gypsum is the main component of joint compound used for drywall construction and is sometimes applied as a skim coat over wallboard.

Today, a number of straw bale builders in the United States, Canada, and Australia are finishing interior walls with gypsum plaster. As you will soon see, gypsum can be applied directly to straw bales or over an earthen base coat on straw bale walls. Gypsum plaster can also be applied to earthen walls, such as adobe and cob, either directly or over an earthen plaster base coat for interior applications.

9-1: Gypsum-based joint compound and texturizing compound are now widely used in conjuction with the gypsum-based drywall, which is used in standard construction. As shown here, gypsum plaster is frequently found in older homes, applied over wooden lath — horizontal pieces of wood that serve as an attachment substrate.

What is Gypsum Plaster?

Gypsum is a naturally occurring water-soluble mineral known by chemists as *hydrous calcium sulfate*, which is found frequently as well-formed crystals. Some of these can be very large: single crystals with a diameter of six feet have been unearthed in Naica, Mexico. Widely distributed throughout the Earth's surface, gypsum is found in rather thick deposits among layers of shale and limestone and under salt deposits in the Earth's crust that were formed as ancient seas evaporated. In addition, gypsum is found in volcanic areas and in veins of metallic ores.

Gypsum plaster is made by heating gypsum at low temperatures to drive off water, to produce *plaster of Paris* — a chalky, fast-setting material commonly used for molds and (less commonly these days) for casts to repair broken bones. It can be mixed with lime to form a fast-drying interior finish plaster known as gauging plaster, according to Bob Campbell, a 17-year plasterer who began work in Scotland. Lime and gauging plaster, he tells us, aren't very popular any more.

If gypsum is heated at higher temperatures, the resulting product is anhydrous calcium sulfate. This substance is most commonly used to manufacture drywall and drywall compound. It is also used to as an interior plaster.

Natural gypsum is by far the best plaster on the market. However, in its pure form it may be difficult to locate. Gypsum plaster is manufactured by several large companies — Red Top, Structolite, Durabond, and Sheetrock 90 are common brand names. Gypsum plaster is available in two forms: wet (in buckets) and dry or powdered form (in bags). Neither option contains sand, as sand is not commonly added to gypsum plaster — at least not in North America. In Australia, however, Gary Dorn and his colleagues routinely add sand to gypsum plaster. (More on this later.)

As you study the various gypsum plaster products, you will find that manufacturers add various chemicals to accelerate the set time and to ensure a harder plaster. Premixed wet plasters in buckets contain a lot more chemicals (binders, drying agents, and preservatives) than do dry powders, which may pose a potential health risk for some people. Although the dry powders contain fewer chemicals, they still contain potentially harmful substances that could affect your health. To avoid exposure to such substances, during and after application (due to release of chemicals as vapor), we strongly recommend that you use a product that does not contain chemical additives, one that is also asbestos-free.

FLY ASH GYPSUM

Some manufacturers are now producing a "synthetic" gypsum for use in drywall, joint and texture compounds, and gypsum plasters. Synthetic gypsum is plaster made from a hazardous waste material, known as fly ash, collected from air pollution control devices on coal-fired power plants. Fly ash contains potentially toxic heavy metals. Although industry officials say that the levels won't cause any harm, we prefer a cautionary approach in such instances and recommend avoiding the use of this product.

If you decide to use standard off-the-shelf preparations, be sure to do so in well-ventilated spaces, and allow some time for the products to release vapor before anyone moves in. If you are chemically sensitive, commercially available gypsum plasters will very likely not be suitable.

When gypsum plaster dries, it produces a hard, durable finish that can last many years. However, it is softer and more water absorbent than lime plaster, and will wash away quite readily if exposed to rain; as a result, it is only recommended for use on interior walls.

As Chris Magwood and Peter Mack point out in their book, *Straw Bale Building*, when gypsum plaster is used indoors, it "can last the lifetime of your home with no maintenance required." They go on to say, "Anybody who has ever stripped an old home of its plaster will know how durable it can be!" However, on an exterior wall exposed to the rain, gypsum plaster wouldn't last very long: it is just too water soluble.

JOHNNY WEISS

9-2: Gypsum plaster produces a durable finish coat with very little cracking. Here gypsum plaster is applied directly over adobe walls.

Why Use Gypsum Plaster?

Many commonly used gypsum plaster products are widely available in North America; they can be purchased from almost any lumberyard, building supply outlet, or hardware store. They are also widely available in other parts of the world. If a local outlet doesn't stock gypsum plaster, they can certainly order it for you.

Gypsum plaster is extremely easy to mix — much easier than any other type of plaster you'll encounter. You won't be mixing separate ingredients as you would when preparing an earthen or lime plaster, though bagged gypsum plaster requires the addition of water. After thorough stirring, you are ready to begin work. (As noted above, sand is not typically added to a gypsum plaster.)

Gypsum plaster is also relatively easy to work with. It doesn't require much skill to mix or to apply to straw bale or earthen walls — as long as you're not trying to achieve a perfectly flat, smooth finish. To achieve a flat surface, you may need to use a four-foot-long screed trowel, which will require some practice (Figure 9-3). You might consider hiring a professional.

> ### INTERIOR USE ONLY!
> Gypsum plaster produces a hard, durable finish that can last many years, but as plasters go, gypsum is rather soft and water absorbent material. It can only be used for interior walls.

Gypsum plaster is light, very sticky, and pliable (or plastic) material when wet. As a result, it adheres extremely well to almost anything, including straw bales, earthen plaster,

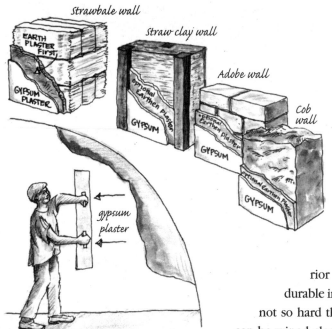

strawbale wall

straw clay wall

Adobe wall

Cob wall

EARTH PLASTER FIRST

GYPSUM PLASTER

OPTIONAL EARTHEN PLASTER

GYPSUM

Optional Earthen plaster

GYPSUM

Optional Earthen Plaster

GYPSUM

gypsum plaster

9-3: Creating a truly flat surface with gypsum plaster requires skill and patience and lots of experience. For newcomers, it might be preferable to hire a professional. Usually gypsum is applied by trowel and sometimes finished using a "screed" trowel, as shown here.

and earthen wall materials such as cob and adobe. No binding agents need to be added to make it stick, either. Furthermore, gypsum plaster adheres well to straw bales no matter how they're laid — flat or on edge — and no stucco netting is required. "Plastering is easier with gypsum than with any other material. Upside down, sideways, it doesn't matter. It sticks," says Magwood. On this topic, Paul Lacinski and Michel Bergeron note: "Applying gypsum plaster is about as much work as frosting a cake."

Gypsum plaster can be applied in many different ways to achieve a variety of visual effects. It can even be sprayed on walls using a drywall texturing gun, and stucco pumps are sometimes used.

Gypsum plaster can create beautiful finishes for interior walls. Although it is light, it sets quite hard — producing a durable interior wall and ceiling finish, though not as hard as lime and not so hard that it can be used for exterior walls. Most gypsum plasters can be wiped clean if soiled and are not subject to dusting or crumbling. Be sure to use gypsum plaster for a finish coat with a hard, durable surface which requires little maintenance. Structural stress will cause it to crack, but this is true for any plaster.

Although gypsum produces a hard, durable finish, it is softer than lime plaster and cement stucco and therefore acoustically superior. Gypsum-covered walls perform acoustically more like earthen plastered interior walls.

Another advantage of gypsum plaster is that it safe to work with, much like earthen plaster. It is certainly much safer than lime plaster: it does dry and crack the hands a little, but is not caustic like cement or lime. Like all other types of plaster described in the book, gypsum is naturally fire resistant — a good attribute for straw bale homes.

Yet another benefit is that gypsum plaster does not shrink upon drying: in fact, it may actually expand a bit as it sets. Consequently, cracking is generally not a problem. Hairline cracks self-heal when moistened — that is, if you daub a little moisture on a crack, it will seal.

As noted earlier, gypsum plaster can be applied over an earthen plaster base coat that keys into straw bales and evens out irregularities in the wall's surface. For straw bale homes, you'll need to apply a clay slip coat first, then the discovery and infill coats. When they've dried, you can apply a gypsum top coat, which allows you to use low-cost, locally harvested, natural materials and to minimize the use of gypsum, a material with a much greater environmental impact. Gypsum plaster can be applied over cement stucco base coats on straw bale walls, too, although we don't recommend this practice because of moisture problems — that is, because cement stucco traps moisture inside straw and earthen walls, leading to mold and structural damage.

Gypsum plaster can be sanded to remove high spots and then pigmented or painted. Sheetrock 90, one commonly used product, dries bright white. Structolite, another commonly used product, cures to a dull gray or pink, which can then be primed and painted to achieve a more appealing wall color. (In natural homes, be sure to use a breathable paint.) The color of gypsum plasters results from clay "contamination" of the deposit from which the material was extracted.

One of gypsum's most important attributes is that it is highly vapor permeable or breathable. Gypsum is also suitable for molding or carving to create relief elements in walls, though the skills needed to create ornate moldings require a tremendous amount of training.

Gypsum plaster is easy to repair. If your tastes run along more conventional lines — that is, you are looking for interior walls with a drywall look — gypsum plaster may be the product for you.

Several gypsum plaster products are available, each with specific applications. For a base coat, some natural builders recommend Structolite, which is bulky and goes on rather quickly. Sheetrock 90, on the other hand, they say, makes a good finish coat. Once again, we recommend a natural gypsum plaster for straw bale and other applications to avoid exposure to potentially harmful toxic chemical additives.

Gypsum plaster is manufactured under fairly careful conditions, so you're ensured a consistent, good-quality product. The tangible benefit? No experimentation will be required to get the mix right — as is necessary in all other forms of plaster we have discussed. You may want to experiment with different products, however.

Gypsum is code-approved, meaning you should have few, if any, problems getting approval for its use. Building code officials know about gypsum plaster. Companies that supply the material can also be called upon to obtain information on strength and durability, should code officials have any questions or doubts about its suitability.

Another benefit of gypsum for professional plasterers or experienced nonprofessionals is its fast set time. Writing in *The Last Straw*, Gary Dorn of Perth, Australia, notes that the fast set time — ranging from 30 minutes to two hours — means that subsequent coats can be applied immediately. This, says Dorn, means that "you can apply two or three coats of gypsum plaster to internal walls of a building within a few days and move in, without having to wait months for the drying process, like you do with lime plasters."

(For earthen plaster, the wait time ranges from a few days to a couple of weeks, depending on climate.)

Although gypsum plaster has many advantages and may work well for interior plaster applications on natural homes, it does have some downsides.

Disadvantages of Gypsum Plaster

One problem with gypsum plasters is the potential for release of chemicals into the air. Binders, fungicides, setting agents, and other chemicals may adversely affect indoor air quality for several years. Applicators should work in a well-ventilated space or use a natural gypsum product, if available. Those who are chemically sensitive and those concerned about potential health effects should research options that will allow them to avoid the use of products that could pose a problem.

Another downside of gypsum plasters is that they have a higher embodied energy than earthen plasters. These product are mined, processed, and manufactured a long way from the job site. A considerable amount of energy is used both in processing gypsum and in transporting it to market. In order of embodied energy, the plasters stack up as follows: earthen plaster has by far the lowest embodied energy; gypsum is next; then lime; and cement stucco has the highest of all.

Gypsum plaster sets up fast — very fast: between 30 and 120 minutes, which can be a problem for the novice. Sheetrock 90, for instance, has a 90-minute set time. Structolite sets up more quickly — a little more than an hour. Don't be lulled into complacency, however: a 60-to-90-minute set time does not mean you have 60 to 90 minutes from the time of application to get it right. The clock starts ticking the minute the material is mixed with water. In addition, there's a point before it has set when it is no longer workable. Sheetrock 90, for instance, can be worked for only about an hour after you've mixed it, although it takes another 30 minutes to harden. Put another way, Sheetrock 90 is fairly hard an hour after you start mixing, after which time it is pretty much unworkable. This creates time pressure, especially for inexperienced plasterers.

Gypsum plaster can harden on tools, too, creating a serious cleaning problem. If not cleaned up quickly, tools can be ruined by gypsum plaster. Gypsum plaster may also harden in mixing containers and buckets, making the plaster unusable and ruining the containers.

GYPSUM-SAND PLASTER

Although in North America, few, if any, plasterers add sand to their gypsum, Frank Warren and Gary Dorn of Perth, Australia have experimented with gypsum-sand-lime plasters. They add three parts sand to one part pre-mixed lime plaster (3 sand: 1 lime) to one part casting plaster and pearl glue, which goes on well, but sets up very quickly — in 30 minutes. They've also worked with five parts sand to two parts casting plaster and pearl glue; this mix went on well, too, but set up more slowly. Both mixes dried fast enough that they could be scratched or roughened almost immediately afterwards in preparation for the final coat.

Source: Gary Dorn, "Gypsum Plaster on Straw Bales," *The Last Straw*, No. 33, p. 28

Because of its fast set time, a skilled applicator generally mixes only about a half a bag at a time. As a neophyte, you will very likely need to mix even smaller quantities at first. Once you've become adept at it, you can mix more at a time. As a rule, when you are just starting out, don't mix any more than you can apply in 10 to 15 minutes. We recommend cleaning tools between batches.

Finally, on the downside, gypsum plaster is fairly expensive. You'll pay two to three times more for it than for lime plaster, and even more than for an earthen plaster — even though gypsum plaster is a widely available product.

Mixing Gypsum Plaster

CHRIS MAGWOOD

As far as plasters go, this is truly the easiest of all products to mix. Gypsum plaster purchased in powdered form is mixed with water and troweled onto walls; wet (premixed) gypsum can be troweled on right out of the bucket. With either option, there's generally no need to add sand unless you want to add some inexpensive bulking agent to the base coat. If you decide to add sand, be sure it is clean and well screened, and run it through a ⅛-inch hardware to remove debris and large grains. For finer coats, you may want to run the sand through a metal window screen. Add two to three parts sand to one part gypsum powder. Adding sand to the mix, to increase bulk in a base coat, makes it set up more quickly.

Gypsum plaster can be mixed in clean buckets, wheelbarrows, 50-gallon drums cut in half, or cement mixers. The drum method, used by Chris Magwood and his partners, results in less splashing than wheelbarrow mixing. They use a heavy paddle mixer on a ½-inch drill for efficiency. For big jobs, they place their barrels on dollies so they can be rolled to applicators (Figure 9-4).

9-4: *Half barrels (shown here) are great for mixing and applying gypsum plaster. Placing half barrels containing gypsum plaster on dollies makes it easier to transport the plaster around work areas.*

When mixing gypsum plaster, be sure not to add too much water, which can make the mix runny and more difficult to apply. In addition, when the water evaporates it may leave empty spaces that weaken the plaster, and the surface may appear pockmarked.

Also, when mixing plaster, be sure to use clean water — as free of soluble salts as possible. Otherwise, these salts will be drawn out of the plaster over time by water vapor and deposited on the surface, creating an ugly problem known as *efflorescence,* which can weaken the surface and cause spalling. Although cracking is less of a problem than with other forms of plaster, some veteran plasterers add fibers such as horse hair, jute, sisal, or straw to reduce cracking. Fine hairs or fluff from cattails may work as well. Fibers are best for base coats and should be kept out of finish coats.

Though gypsum sets up quickly, there are ways to slow the set time. For instance, you can use very cold water to mix gypsum. A number of different substances such as sodium citrate, lactose, maltose, and sucrose also slow the set time, giving you more time to work. If you are adding lactose, maltose, or sucrose, you will need to add a small amount of lime to the plaster. Because all of this can become quite complicated, the beginner should probably just make small batches of gypsum plaster at the outset, and learn to work fast.

According to Athena and Bill Steen and co-authors of *The Straw Bale House*, one of the most versatile all-around gypsum plaster products is Red Top, which can be purchased in regular and fibered mixes. Red Top is easy to work with and bonds well to earthen plasters, they note. Chris Magwood typically uses Sheetrock 90 and Structolite. You may want to purchase several gypsum products, especially natural gypsums, and try each one before settling on one brand.

Applying Gypsum Plaster

As noted earlier, in natural homes gypsum plaster works well when applied over an earthen base coat. When plastering straw bales, we recommend that you apply earthen plasters over the straw bales to smooth out the wall. You'll need to apply a slip coat first, then a scratch coat, followed by an infill coat: gypsum plaster goes on next. If your wall needs further smoothing out, you may need to apply a second coat of gypsum. Remember: straw bale walls should be trimmed and shaved first, as explained in Chapter 3.

9-5: Gypsum plaster is so adhesive that it can be applied directly on to bales.

CHRIS MAGWOOD

When applying gypsum over a dry earthen surface, wet the wall first so it doesn't suck the water out of the newly applied plaster, which will cause the gypsum to flake off. By wetting the wall first, you'll create a strong bond between the gypsum and underlying earthen materials.

Gypsum plaster can also be applied directly to straw bale walls and is typically applied by trowel, although it may be applied by hand as well (Figure 9-5). As Frank Warren, a master plasterer from Australia puts it, gypsum plaster will stick "pretty bloody well." Moisten the bale surfaces before applying a gypsum plaster to them, too.

Before you begin work, test your plaster and your technique on a small section of a wall or a spare straw bale or two. Apply plaster over an area about one- or two-feet square to see how the mix "performs" — that is, how easily it goes on, how well it keys in, and how quickly it dries. If no large cracks appear, it is probably okay to begin work. If cracks appear, you may want to try a different gypsum plaster product.

Gypsum can also be applied directly on bales and earthen walls: there's no need to attach chicken wire. Like other forms of plaster, gypsum is applied in successive coats, but typically only two coats, a rough coat and a finish coat, are needed on earthen walls. The rough coat is about ¾-inch thick; the finish coat, which is usually applied a day or two later, is about ¹⁄₁₆- to ³⁄₈-inch thick.

When applying gypsum directly on straw bales, Chris Magwood and his business partners usually apply three coats: two coats of Structolite and a finish coat of Sheetrock 90. They use trowels for the first two coats, as hand application doesn't work well; gypsum sticks just as well to skin as it does to bales, making it hard to apply this way. The finish coat can be applied by trowel, roller, or brush.

As Gary Dorn notes, when working with straw bale, it is worthwhile to fill all holes with a gypsum plaster mix before applying the scratch coat. Frank Warren recommends a mix consisting of two parts casting plaster (plaster of Paris) to five parts sand with some skim milk powder (as a set retardant) to fill holes and cracks. The scratch coat will go on more smoothly as a result.

When using a trowel on a finish coat, a smooth surface can be produced by thinning the plaster with a little water, then brushing it on. Rollers and brushes allow the wall to be texturized. Rollers, for instance, give a typical "stucco" look with tiny bumps — that is, a grainy look — while brushes can be swirled or daubed onto the wall to create a variety of textured surfaces. A float sponge or whisk broom can be run over the surface to give it texture, too. Adding sand to the finish coat gives it a grainy texture. You can also texturize the surface using a drywall texture gun.

For optimal results, be sure to maintain indoor temperatures around 55° F or higher until the plaster has dried. You may need to mist the wall periodically to prevent it from setting too quickly, especially in drier climates.

Painting and Sealing Gypsum Plaster Walls

Gypsum plasters dry to various shades of white, and occasionally to a gray, pink, or peach color, depending on the brand. As a result, many people pigment or paint gypsum walls to provide color (see color gallery).

As in earthen plasters and lime plasters, pigments can be added to the finish coat. They are typically mixed with water in a jar, then added to the dry gypsum, or mixed in powder form with the dry gypsum plaster powder. After the two powders are thoroughly mixed, water is added. When pigmenting gypsum, test the mix (apply it and let it dry) to be certain it achieves the desired color and that the pigment does not compromise the structural integrity of the plaster.

Gypsum plastered walls can become dirty, so many builders seal walls with paint. As noted above, gypsum can be painted with natural or latex paints. Although latex paints are widely available and are used in conventional construction over drywall, we strongly recommend against their use in homes built from straw or earth. Standard latex paints pose a health risk (as a result of the release of toxic chemicals) and also reduce vapor permeability or breathability of walls. Natural paints, such as the milk-based paints, are a better choice from a health standpoint and they'll allow your walls to breathe. (See Chapter 7 for a detailed discussion of natural paints.) If you're going to apply a natural paint or an alis over gypsum plaster, be sure that the gypsum plaster product you've selected allows the paints or alises to adhere well.

> ## CREATING SPECIAL EFFECTS
>
> Some applicators like to use a paint brush to splatter colors (water-borne pigments) on the wall over freshly applied plaster. The wall is then troweled smooth. The result is a mottled look that can be quite stunning.

While we're on the subject, natural paints can be used to seal gypsum walls, which can be dusty. Oil can also be used to seal gypsum plaster walls, and some builders use natural linseed oil. One part linseed oil is mixed with one-half part gum turpentine, then rubbed onto the surface of the wall. For more natural and healthier options, use oil sealants from Bioshield, OS Color, Auro, AFM, and Livos. Natural oils can be sprayed onto the surface of a wall, which are then wiped down to remove excess. Be aware, however, that oil can yellow a wall. Hairline cracks in the wall will darken, creating an antique appearance. Walls can also be buffed with natural beeswax finishes or natural oil-based wall glazes. See Chapter 7 for a discussion of natural wall finishes.

Quick Recap

Gypsum plaster is made from a naturally occurring, water-soluble mineral and has been used for centuries, but is limited to interiors. Unlike other plasters, gypsum generally does not contain sand but typically contains several chemical additives to accelerate set time and to ensure a harder plaster. Health problems may result from inhalation of these substances during application and after plaster has been applied. So when using gypsum be sure to work in a well-ventilated space and let the material dry completely for several weeks before moving in.

Despite this shortcoming, gypsum works well on earthen or straw bale walls, or over an earthen base coat. Gypsum plaster goes on easily and dries quickly, forming a hard, durable finish that can last for many years. The finish coat can be pigmented or painted to achieve the desired color.

CHAPTER 10

Plastering Earthen Walls

NATURAL BUILDING encompasses a wide range of materials and techniques, providing each of us an opportunity to create shelter ideally suited to the local climate, out of locally available materials. As noted in Chapter 1, natural buildings are typically made out of earth, wood, and fiber. Stone, logs, and even cordwood are also used to construct natural homes. But your options don't end here: a growing number of people are building homes from recycled materials such as tires.

So far, we've focused our attention primarily on plastering straw bale homes, but they're not the only natural or alternative homes that are plastered. In this chapter, we'll focus primarily on earthen structures, including cob, adobe, rammed earth, rammed earth tires, earthbags, and straw-clay. We begin by examining some of the key differences between plastering straw bale homes and earthen homes, and then examine things you need to know to successfully plaster these structures. Be sure to read Chapters 1-9, as they contain lots of information that applies to earthen homes.

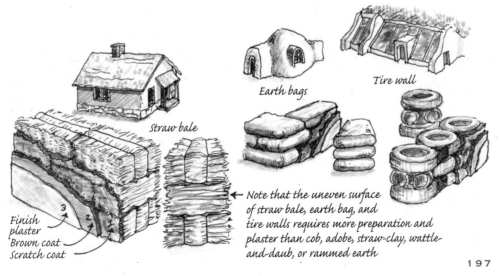

10-1: The amount of wall preparation depends on the material or building system. Straw bale, earthbag, and rammed earth tire all require additional preparation.

Tire wall

Earth bags

Straw bale

Finish plaster
Brown coat
Scratch coat

← Note that the uneven surface of straw bale, earth bag, and tire walls requires more preparation and plaster than cob, adobe, straw-clay, wattle-and-daub, or rammed earth

197

10-2: Cob, straw-clay, adobe, wattle-and-daub, and rammed earth walls require minimal wall preparation.

Key Differences Between Plastering Straw Bale Walls and Earthen Walls

One significant difference between straw bale and most other natural homes is that the latter generally require much less wall preparation prior to plastering. In straw bale construction, you must first "shave" bales, generally with a weed eater (or some similar tool), to shape walls and remove loose straw, to ensure an even finish and a tighter bond between the plaster and the underlying straw bales. After this, you spray on a slip coat to key your earthen plaster well into the bales, and then pack straw and a straw-mud mix into joints between bales. With this work completed, plastering begins. In contrast, plastering most earthen walls requires much less prep work — in most cases, it begins as soon as the wall is erect and dry (Figures 10-1, 10-2).

Another difference between straw bale and other earthen natural building materials is that, for the most part, earthen wall systems, such as those made from adobe or cob, generally require only two coats of plaster: a scratch coat and a finish coat. In some instances, you may choose not to apply a plaster at all. For example, cob homes built in the Willamette Valley of northwestern Oregon, home of the Cob Cottage Company, are typically left unplastered. Although it rains a great deal in this part of the country, rain generally falls straight down; cob walls don't need a protective layer of plaster as long as they're protected by good overhangs. In some instances, a finish coat is applied — for protective or decorative purposes.

In contrast, Michael Smith writes, in his book, *The Cobber's Companion,* "On the west coasts of England and Wales, cob buildings are blasted by frequent North Atlantic gales, which drive rain horizontally at great speeds." In this area, a lime-sand plaster is applied, which holds up well in this harsh climate. In such instances, design features (discussed in Chapter 3) are crucial in protecting a lime-sand plaster, as important as they are in protecting an earthen plaster.

Although a protective coat of plaster may be omitted, plastering earthen homes guarantees a longer life. After plastering, alisés can be applied to provide additional protection. The formulas for these plasters and finish coats are presented in Chapter 6. Because subsoils vary with respect to clay, silt, and sand content from site to site and even on a given building site, you'll have to test mixes to see what works best for your project. You can also talk to experienced natural plasterers in your area who may be able to provide you with some guidance. They could, for instance, give you advice on recipes that could serve as a good starting point for your tests.

Another difference between straw bale and some earthen homes is that in the latter, plaster tends to go on more easily. For example, straw-clay walls present a fairly flat, even surface for plastering. As a result, you generally won't have many dips and valleys to fill in or build up as you would with a straw bale home.

Because there's less wall preparation and fewer coats to apply, and because plaster goes on relatively easily, plastering most earthen homes requires less time and less effort than plastering a straw bale structure.

With this brief comparison in mind, we turn our attention to several traditional types of earthen homes — notably, cob, adobe, rammed earth, straw clay, and wattle and daub — and two alternative building techniques, earthbags and rammed earth tire homes (such as Earthships). No matter what system you're using, be sure to check out what experienced natural builders have to say about plastering. We also recommend that you read books and articles and attend a natural building workshop or two on your preferred building technique. Learn as much as you can about natural building and plastering earthen walls and experiment with natural plastering *before* you begin your project and be sure to follow the design guidelines presented in Chapter 3.

Just Say No to Cement Stucco

Before we examine the various types of plaster, permit us another word on cement stucco: applying cement stucco to earthen walls will almost certainly spell disaster for natural and most alternative buildings — in all climates! We strongly advise against the

CEDAR ROSE GUELBERTH

10-3: Cement stucco is a bad choice for earthen buildings. It does not adhere well to earthen surfaces, and it wicks moisture in toward the earthen walls. Because cement-based stucco and earthen materials expand and contract at different rates, cement stucco tends to crack — allowing moisture into the wall, which can cause serious structural damage. Here the cement stucco has separated and fallen off, exposing the inner adobe blocks.

10-4: This adobe house is distintegrating as a result of the cement stucco finish that allowed water to penetrate the surface and destroy the underlying adobe brick walls.

use of cement stucco, in spite of the fact that until recently, New Mexico code required it for adobe homes. Why?

There are several reasons. First, cement does not bond well to earthen materials, even if applied over wire mesh or lath, as described in Chapter 2. (Remember: like materials adhere best to like materials and earth and cement are not "like" materials!) Second, cement-based stucco and earthen materials expand and contract at different rates. The result? Cement stucco tends to crack. As we noted in Chapter 2, cracks in cement stucco are rather hard to repair.

Cracking of cement stucco allows moisture to enter the interior of the wall. More importantly, cement wicks moisture, drawing it into the earthen wall. Because cement stucco won't allow water vapor to escape, moisture accumulates inside walls, causing earthen materials to dissolve. While the cement stucco facade may look fine, the interior wall damage may not be apparent until it has reached a critical stage. At this point, the deterioration of the earthen wall can cause serious structural damage, even the collapse of a building.

This warning isn't based on speculation or the result of a natural builder's prejudice against cement stucco. Over the years, many adobe buildings in the United States and

CEDAR ROSE GUELBERTH

numerous cob buildings in England and New Zealand have been seriously damaged by applying cement stucco to exterior walls. Some buildings actually collapsed as a result of this well-intentioned application of cement stucco (Figure 10-4).

Plastering Cob Buildings

Cob is a mixture of clay, sand, and straw, and is used to construct walls by hand. When cob is dry, it becomes as hard as sandstone. This technique allows the construction of creative free-form sculpted walls (Figure 10-5). As noted earlier, exterior walls can be left as is, or they can be plastered to provide protection and/or for aesthetic reasons. Choose your materials carefully: you will need a plaster that is durable and breathable — that is, it permits water vapor to pass through.

Interior Plasters

While plastering an exterior wall in a cob home is optional, interior plasters are frequently desirable. Because cob is always the color of your dirt, you may want to consider plastering interior walls with a pigmented mix to lighten walls and provide color. A smooth, light-colored finish adds considerable beauty to a home. Interior plasters also reduce dusting and are easier to clean than unfinished cob, which has a rough surface.

For interior walls, you can use three types of plaster: earthen, lime, and gypsum plaster. All are breathable and quite durable as interior plasters. Many cob builders prefer earthen plasters and earthen finish coats. Finish coats can be pigmented for color or coated with an alis, a clay finish coat, lime plaster and limewash, or a pigmented casein paint.

IANTO EVANS · COB COTTAGE COMPANY

10-5: Ianto Evans forms a cob wall. Sculpting and shaping a wall and working to provide an even surface during construction can make your wall prep and plastering an easier task.

Many natural builders apply lime plaster finishes either on earthen walls or over earthen plaster base coats (see color gallery). Lime plaster is a good choice, as it expands and contracts at the same rate as earthen materials, resulting in a stable plaster application. Lime plaster produces a stunning white finish to which pigments can be added for color, and gives a very durable finish, making it especially useful for kitchens and bathrooms, as it deters mold growth. Remember that good ventilation is essential in any high-humidity room — or any home, for that matter. Gypsum plaster can be applied to interiors, as well (Chapter 9).

Even if you've decided to finish your interior walls with gypsum or lime, you may want to apply an earthen base coat first. Earthen plaster can be used to create a relatively smooth and attractive surface over which the finish coat can be applied. Earthen base coats are easy to apply and inexpensive, as well. Pigmenting a lime or gypsum finish coat adds color to a room (see Chapter 7). Lime whitewash can be applied to lime plaster walls (discussed in Chapter 8).

While we're on the subject of wall finishes, be careful to avoid commercial oil or latex paint, which are not breathable and thus do not allow trapped moisture to escape, which could cause serious damage to your walls. Select breathable, vapor permeable finishes such as lime washes, silica and casein paints, and natural oil or beeswax finishes on plaster-finished cob walls (described in Chapter 7).

Exterior Plaster on Cob Buildings

The exterior walls of cob buildings are plastered with either earthen plaster, lime plaster, or litema. Earthen plasters go on easily and can be used for infilling and shaping walls — and they provide a beautiful protective finish. Lime-sand plaster is generally used in wetter climates where driving rain is a problem, although with proper design (discussed in Chapter 3), a builder can apply an exterior earthen plaster in some rather nasty climates. As noted in Chapter 8, exterior lime-sand plaster can be applied directly to cob or as a finish coat over an earthen plaster base coat.

Lime plaster can be applied either over an earthen plaster base, as just described, or directly to cob walls. In the latter case, you'll need to use a cob-like mix to infill and shape the wall, then apply lime plaster (as described in Chapter 8).

Litema is composed of colored clay and fresh animal manure. Smeared on the walls, it provides a durable and delightfully colorful surface (discussed in Chapter 7).

Wall Preparation

Although most natural homes — other than straw bale structures, that is — require little advance preparation, there is some work that needs to be done before plastering a cob wall. First, while building your walls, be sure to trim off any excess cob that protrudes from the walls. As much as possible, shape and trim cob walls while you are building them — that is, while they are still moist. You will save a lot of time and effort

10-6: Irregularities in a cob wall can be trimmed off by using a machete or other similar tool, such as this home-made sharpened scraper formed from a shovel.

if you do this prep work while the wall is still moist rather than when it is rock hard. Cob builders typically use a machete or a spud — a sawed-off, sharpened shovel (Figure 10-6). Hand saws, hatchets, and axes can also be used to trim cob walls in preparation for plaster.

Be sure the walls are completely dry before applying a plaster. If there are deep depressions in the cob, you will need to fill them with an infill layer (described in Chapter 6) *before* you begin plastering, as in a straw bale home. Use a fairly moist, sticky cob or infill plaster mix in this instance, containing straw of many different lengths and a bit more clay than the mix used to make your walls. Note, too, that you may need to roughen the surface of any depression to increase the adhesion of the cob filler. If necessary, you can carve it out a bit with a sharp implement, such as a hatchet or a knife. Moisten the area well, then allow the water to penetrate the surface, before applying the infill plaster — and, once again, don't wet the walls so much that water runs down them. Remember to feather the infill plaster (described in Chapter 6).

For advice on shaping or sculpting walls or filling deeper depressions, use the infill guidelines in Chapter 6.

Mixing and Applying Earthen Plaster on a Cob Wall

After your walls are prepared, it is time to mix your earthen plaster. Most cob plastering jobs require only two coats. The first coat consists of the same components you used to make the cob — clay-dirt, sand, and straw. However, the mix should be more adhesive, smoother, and wetter than the cob used to build your walls. To achieve a smoother mix, you need to screen the soil to remove small pebbles and rocks: a ½-inch screen will generally suffice. If the sand you're using is coarse and contains large pebbles or stones, you will need to screen it using a ⅛-inch screen.

To begin, see the guidelines for developing plasters in Chapter 6, which serve as a starting point for making plaster. However, as we've said before, bear in mind that subsoils vary from one place to the next, so it is impossible for us — or anyone — to provide a fail-safe recipe for your initial coat. We recommend that you start with the guidelines we've provided for the discovery (and infill) coats in Chapter 6, and work from there. You may want to talk to some experienced mudders in your area and get their views as well. It's not a bad idea to hire a professional or an experienced mudder to help you for a few days: he or she can get you started, help you make a good plaster mix, and show you how to apply it.

After you screen the clay-dirt, add sand, chopped straw (strands from two to six inches in length), and water. Flour paste and/or manure can be added, if needed. Mix the components, as described in Chapter 5. The plaster should be wet enough so that it can be troweled on to the wall or applied by hand easily. Too wet, and the plaster may be difficult to apply; too dry, and it will not adhere well to the surface and will peel off the wall. Your goal is to produce an adhesive plaster that will bind well to the wall, yet with enough structure to shape and fill in the wall.

The adhesiveness of a mix is determined by its clay and water content. Ideally, what you want is a plaster that sticks better to itself than to your hands. Cob builders like to toss a handful of plaster on a wall from a distance of several yards: most of it should adhere to the wall. However, a mix that is too sticky will glom onto your fingers because it contains too much clay and/or water. A little more sand will often help, as will a little less water. If it isn't sticky enough, you may need to add more clay or additives and/or more water. Be sure you don't add too much water, or your mix may become too soupy. If so, add a bit more of the dry components (dirt, sand, and straw). After you test how well the plaster goes on your walls, let it dry to check for cracking, hardness, and adhesion, as described in Chapter 6.

After you've come up with a good mix, it is time to start plastering. You will need to wet the walls first, with a hose or a brush, or daub water on with a large sponge (this, however, is time-consuming and not very effective). When wetting a wall, avoid applying so much water that it runs down the wall: too much water will wash away the clay. However, you'll need to apply enough water to provide even penetration of the surface, in order to ensure a good adhesion of plaster to the wall.

The first coat of plaster on a cob structure is applied by hand or trowel. (For more on applying plaster, see Chapter 6.) It should be about ½-inch thick, and should fill holes and other depressions in the wall, producing the surface you want — which is best accomplished by hand application. Holes and depressions may require application of additional plaster. If applying plaster bare-handed or by trowel, you may want to have a bucket of water to rinse them. Be sure to key the plaster into the surface well. If you use a trowel, don't trowel the surface too smooth; the finish coat, which is applied next, needs a slightly rough surface to key into.

The finish coat is applied over the base coat — but only after it has dried completely. Chapter 6 provides guidelines for finish coats on straw bale walls that apply to cob walls and all other natural homes. Be sure to test your mix before you begin applying the finish coat. The regional variability of the soils could require some doctoring to achieve the proper mix. Remember, when making a finish coat, screen your soil using ⅛-inch wire mesh to eliminate stones and debris that make plastering more difficult.

After you've come up with a suitable finish coat, moisten the wall (the surface of the base coat) before you apply your finish coat either by hand or by trowel. Finish plaster is usually applied about ⅛- to ¼-inch thick and one coat is usually sufficient. However, an additional finish coat may be needed to achieve a truly smooth finish — if that's what you are looking for. Read Chapter 6 for a discussion of ways to manipulate the texture and the color of the finish coat.

Remember that trowels tend to give a smoother, harder, and more durable finish. Hand application tends to produce a rougher finish coat. Once it has set up, an earthen plaster can be buffed with a piece of damp burlap or a sponge, using broad, circular motions to add texture and to remove any irregularities in the surface, such as large grains of sand. Alternatively, once it has set up, the finish coat can be polished to close the pores in the surface, using a plastic disk, such as a yogurt container lid (as described in Chapter 6), or a hard trowel.

Be sure to avoid application in direct sun or during cold weather. Time your application so that it doesn't dry too fast. If it does, the plaster may crack excessively. (Remember, though, cracking is also a sign that the mix doesn't have enough sand or fiber in it.)

Plastering Adobe Walls

Adobe homes are made from adobe blocks or bricks, traditionally made by hand from a clay-rich soil with sand and straw. The mud is mixed, then poured into wooden block forms and left to bake in the sun (Figure 10-7). The dried blocks are then laid up in a running bond, using a mud or earthen mortar made from the same material you use to produce the blocks, although screened to remove pebbles. After the blocks are made, they're laid on a foundation, one course at a time, in a running bond pattern for maximum strength. Mortared in place using the same mud that's used to make blocks, the adobes form an attractive, durable wall.

10-7: Traditional adobe blocks are dried in the sun for a couple of weeks before being used to build walls.

Although most adobe homes are plastered, some are not. Plastering adobe walls helps to ensure a longer life for a building. Cement and mud plaster have both been used on adobe homes. Of the two, earthen plaster is by far the superior choice! It not only protects the wall from the elements, it goes on smoothly, is easy to make, uses locally available materials, and costs very little. Earthen plaster can also be repaired and restored easily.

Preparing Adobe Walls for Plastering

As a general rule, interior and exterior adobe walls need only two coats of earth plaster: a base coat and a finish coat. In addition, adobe walls require little, if any, prep work: it is generally not necessary to trim walls or do anything else to prepare them for natural plaster, especially if the mortar has been applied cleanly — that is, if it doesn't bulge out of the joints past the surface of the walls. If it does, it needs to be leveled out. Be sure that the mortar between your blocks has dried thoroughly before you begin plastering.

DETERIORATION OF ADOBE

Over time, the exterior wall surfaces of unplastered adobe homes will slowly erode as a result of driving rains and wind. If the blocks contain 10 to 20% clay, they generally do a better job of resisting erosion from rain. If the clay content of the blocks is lower and the surface is sandy, rain and wind will do considerably more damage, especially on the side of the building most exposed to prevailing weather. A well-maintained plaster will protect an adobe wall from deterioration.

After the walls have dried, brush them to dislodge and remove loose mortar and adobe on the surface of the bricks. Moisten the surface before applying the base coat: wet the walls well, but not so much that water flows down the surface. Allow the water to penetrate the surface, then apply the plaster mix well to the adobe blocks and mortar.

Plastering Adobe Walls

We recommend two types of plaster for exterior adobe applications: earthen and lime-sand. Earthen plasters and lime-sand plasters work well in a wide range of climates, as long as the walls are protected from weather and as long as one follows the design con-

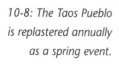

10-8: The Taos Pueblo is replastered annually as a spring event.

CEDAR ROSE GUELBERTH

siderations outlined in Chapter 3. However, if your house design calls for parapet walls, be warned that these are generally only suitable for dry climates. Even then, you will need either to use a lime finish for the entire building, or coat the walls with earthen plaster then cap the parapet walls with lime plaster, or perform yearly maintenance to refinish the earthen plastered parapets. Traditionally, adobe buildings (such as those in pueblos in the Southwest) are re-plastered as part of an annual gathering in the spring, if damage has occurred during the year (Figure 10-8).

To make a plaster for adobe homes, begin by reviewing the discovery and infill plaster sections in Chapter 6. Your mix will be a combination of these two, containing clay-rich dirt, sand, and fiber. As with cob walls, you will want an adhesive mix — that is, a plaster mix with enough clay to stick to the wall. You want a mix that not only adheres well to the wall, but also has some structure to it that will even out the surface.

The base coat should be reinforced with fiber, though too much fiber can make it difficult to produce an even surface. When using straw, use lengths of two to four inches, to provide strength and help prevent cracking. The mix should also contain enough sand to permit the mix to be spread evenly on the wall, to provide structure and fill in depressions, and to prevent cracking. As with all base coats, minimal cracking is okay so long as the plaster binds strongly to the wall.

The first coat can be applied by hand or trowel or a combination of the two. Be sure to key it well into the surface. Don't trowel it smooth, however. The finish coat needs a rough surface to key into. As in cob and other earthen homes, you will find that mud goes on the walls smoothly and easily because the wall surface is already fairly well formed and shaped.

The base coat keys into the adobe wall and creates a strong surface into which the finish coat will adhere. The earthen finish coat is made from screened clay-dirt, sand, and fiber and/or manure. Natural additives are optional. See details on making finish coats in Chapter 6.

Apply the finish coat thinly by hand or with a trowel, then work the finish to achieve the desired texture. As with any plaster, apply the finish coat out of direct sunlight. As a rule, the faster the material dries, the greater the potential for cracking. Small cracks can be repaired by applying a protective clay slip or alis (Chapter 7).

Applying an adhesion coat (discussed in Chapter 6) over wood, metal, or cement that will be plastered offers an additional measure of protection. Writing in *Adobe and Rammed Earth Buildings*, Paul McHenry notes, "As the mud plaster and wood materials in the earth wall seem to be compatible and have about the same coefficients of expansion, waterproofing and lath are not required." The only exception to McHenry's rule is the wood at door jambs. Here, he says, the vibration produced by the closing of doors tends to cause the plaster to crack and to separate from the door jamb. These areas should packed well with a cob-like mix over anchor screws before plastering to produce a stable, solid juncture.

10-9: This adobe building has an interior gypsum plaster finished with an alis, creating extraordinary beauty.

CATHERINE WANEK

Some adobe builders recommend stabilizing earthen plaster with asphalt emulsion — an oily material that is also occasionally added to adobe bricks to make them water resistant. We recommend *against* the use of this product in plaster and adobe blocks, because it releases toxic substances that could harm both workers and the eventual occupants of the building. Why put potentially toxic materials in the walls of a natural home? Besides being toxic, asphalt emulsion in earthen plasters can make future repairs more difficult.

Some adobe builders add Portland cement to stabilize earthen plaster, in an attempt to make it more weather resistant. However, cement can make a mix crack and become susceptible to spalling. If earthen plasters are done well and attention is paid to design details, they can provide a stable, long-lasting, durable finish. If needed, flour paste, cactus juice, or other natural stabilizers can be used. They're safe to work with and highly effective.

INTERIOR PLASTERS. You have a greater number of options for interior plaster than for exterior walls in adobe homes. Three types of plaster can be used in such instances: earthen, gypsum, and lime. No matter what choice you make, earthen plasters are your best options for base coats: they're easy to make and apply and they're economical. You can then apply an earthen finish coat, painted with an alis, or you can use casein and silica paints or a color plaster finish (all discussed in Chapter 7). Or, if you'd like, you might consider applying a lime plaster over an earthen base coat, especially for high-humidity areas such as bathrooms and kitchens. Gypsum plaster can be used for interior walls, as explained in Chapter 9. (Cement stucco shouldn't be used, in our view, under any circumstances.)

When applying lime or gypsum, brush or dust off any loose material before applying plaster. Be certain the surface is free of protrusions, such as mortar sticking out between adobe blocks. Moisten the wall surface and apply the base coat, keying it into the wall. Apply the finish coat over that (Figure 10-9).

Plastering Rammed Earth Walls

Traditionally-built rammed earth homes are made from clay-rich subsoil, which is slightly moistened and rammed between two forms on the foundation. When the walls have been built up, the forms are removed and voila! you have a beautiful smooth-surfaced

10-10: Rammed earth wall construction commences after the foundation has been built. Forms are mounted on the foundation, filled with soil, then compacted, producing solid rocklike walls, as shown here.

JOHNNY WEISS — SOLAR ENERGY INTERNATIONAL

wall (Figure 10-10). Many people choose rammed earth because of its innate beauty — the walls are truly breath taking — and cringe at the idea of covering it with a plaster. Yet there are valid reasons to apply a plaster to a rammed earth wall — for example, plastering exterior walls provides a protective coat that can be restored if weathered. Such an approach prevents layers of the wall itself from possible erosion due to weather. If you build in a neighborhood of conventional homes, your home will fit in better if you plaster the exterior walls.

Preparing Rammed Earth Walls for Plastering

One of the problems with plastering a rammed earth home is that the walls are frequently too smooth to offer a good keying surface for plasters. As a result, you'll need to do one of three things: score the wall before it sets up — that is, as soon as the forms come off; chip or otherwise roughen the wall surface after it has dried; or apply a clay-slip adhesion coat. Because rammed earth walls are quite smooth, you'll very likely need to apply only one or two coats of plaster. Brush walls to remove loose materials, wet the walls, and then apply plaster. Earthen and lime plasters and litema are suitable for exterior and interior finishes. Gypsum is also suitable for interior surfaces. Be sure to apply an adhesion coat over cement or wood bond beams and sup-

JOHNNY WEISS — SOLAR ENERGY INTERNATIONAL

10-11: Smooth surfaces of a rammed earth wall (shown here) can be made suitable for earthen plaster by applying a mixture of clay slip, sand, and flour paste to rough up the finish. Dip a sponge in the mix, then rub it on the surface, and allow to dry. Alternatively, score the surface by rubbing with a circular motion, using a coarse brush which has been dipped in water.

ports. Refer to our recommendations in corresponding chapters on making and applying these materials. Keep in mind that rammed earth walls are so straight and flat that plasters will go on easily and quickly (figure 10-11).

Plastering Straw-clay Walls

Straw-clay walls consist of straw coated with a thin coat of clay slip. Packed between forms attached to posts or vertical framing timber, straw-clay dries to form thick, insulated wall material. Although the amount of clay slip can be increased to provide a wall with greater thermal mass, these walls primarily provide insulation.

Plastering straw-clay walls is a breeze. Robert Laporte, who spearheaded the straw-clay movement in North America, notes in his book, *MoosePrints:*

CEDAR ROSE GUELBERTH

10-12: Straw coated with a fine clay slip is packed into wall forms, creating a thick insulated wall shown here wrapping a timber frame. Because the surface is level, it is quite easy to plaster.

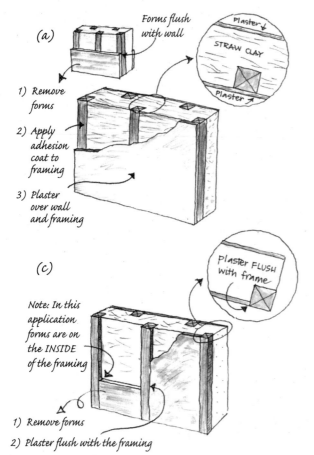

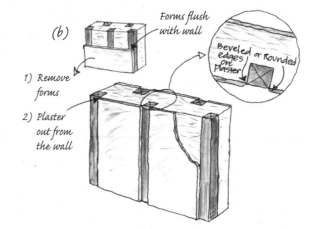

10-13 (a) (b) (c): Three options in straw-clay construction and plastering. Walls can be formed so the straw-clay is flush with the framing. Plaster can be applied over the framing (a), or (b) up to the framing, then rounded or beveled at the edges. If forms are set so the straw-clay wall recedes slightly (c), plaster can be applied flush or inset with the framing members.

A Holistic Home Building Guide, "A straw-clay wall has plenty of texture, and requires no further preparation." Moreover, the plaster goes on quickly and smoothly by hand or by trowel.

Two coats of plaster are all that's required. The base coat consists of a mix of clay-dirt, sand, a minimal amount of chopped straw, and water. Try starting with the scratch coat guidelines we offered in Chapter 6. If you don't have a supply of clay-rich soil, you can supplement your mix with flour paste and manure or add powdered clay. Use the recommendations for a discovery scratch coat in Chapter 6.

Framing members are required for straw-clay construction, which in some cases are covered with plaster. If this is the case, forms should be flush with the frame. The plaster will then be carried over the frame, bridging from one section of the straw-clay wall to the next (Figure 10-13a). Use an adhesion coat over exposed wood to secure this plaster to wood expanses over three inches (described in Chapter 6).

If you want to leave the framing members unplastered, you have two options. You can attach the forms flush to the framing members, then apply the plaster, beveling or rounding it at the edges (Figure 10-13b). Or you can set forms on the inside of the posts, then plaster across the straw-clay wall to the post. Inset the form to insure adequate room for the plaster (Figure 10-13c).

Be sure the straw-clay wall has dried thoroughly before applying the base coat. If you don't, it could slow down drying and cause mold to form in the walls. How do you know your wall is dry? In many instances, straw contains seeds (of the harvest crop) that will sprout in the newly packed wall. If the straw you're using is seed-poor, you can add wheat seeds to it, insuring that plants will sprout out of the walls. When the sprouts wither and die, you'll know the wall is pretty dry — that is, unless you're building extremely thick walls. Bear in mind that drying is a slow process, especially in cold or moist climates. Moisture meters offer a more scientific way to assess the dryness of walls: you can use a surface moisture meter or a unit with a straw probe.

When building with straw-clay, construction should be timed to permit adequate drying — for example, walls should be packed with straw-clay in the late spring or early summer. If you've timed things correctly, you can plaster the wall before cold weather sets in. If not, you may need to wait until the following spring.

Once the forms are removed, you'll find a smooth surface that readily accepts earthen plasters (Figure 10-14). After the wall has dried, wet the surface, then apply the base coat either by hand or trowel; both work well. When working the wall to create a smooth surface, leave the base coat a little rough so the finish coat will adhere more readily or scratch it if it has been troweled smooth.

The first (discovery) coat of plaster is made following directions given in Chapter 6. Be sure not to put too much fiber in this coat, as it can make the plaster more difficult to apply, turning a smooth surface into a lumpy one. After the base coat has been applied, let it dry. As in all other instances, avoid applying plaster in direct sun, to ensure that the walls set up slowly.

After the base coat has dried, it is time to apply the finish coat. Wet it first. Exterior earthen finish coats can be pigmented to add color, or can be coated with a protective alis (Chapter 7). For a highly durable surface in areas with inclement weather (or if desired for aesthetic reasons), lime plaster can also be applied directly to straw-clay or over an earthen base coat.

The finish coat on interior wall surfaces can be earthen, lime, or gypsum — and can be pigmented or left unpigmented. If you want to add color to an earthen finish coat, you can apply a pigmented alis or a natural, breathable paint as described in Chapter 7. Gypsum plaster can also be used to finish interior walls (see Chapter 9). Gypsum plaster is so sticky that it adheres directly to straw-clay walls or over an earthen base coat on straw-clay walls: just wet the wall and apply.

For a really durable interior finish, you can apply a lime-sand plaster. It may be pigmented, frescoed, or coated with a colored lime wash. Lime is quite suitable for high-humidity rooms, for reasons noted in Chapter 8.

LAURA DICKSON

10-14: Straw-clay surfaces (shown here) readily accept earthen plaster, making this an easy application.

PAULA BAKER-LAPORTE

10-15: This straw-clay home designed by Paula Baker-Laporte and built by Robert Laporte is a great example of a beautifully designed, well-built straw-clay home with natural clay plaster finishes.

Plastering Wattle-and-Daub

Wattle-and-daub construction, described briefly in Chapter 1, is rare in North America, but is still widely practiced in many other parts of the world. In fact, there are as many ways of building with wattle-and-daub as there are tribal cultures around the world.

As a rule, wattle-and-daub systems fall into one of two categories. The first is a woven wall system in which flexible branches from trees or bushes, such as willow or wooden lath strips, are woven to create a lattice between framing members (Figure 10.16a). Mudders pack a sticky cob mix onto each side of the lattice, building it out to the desired thickness, which then dries to form a smooth, cob-like exterior over which plaster is applied, usually two to three coats. Because it is a lot like plastering over cob, we refer you to that section in this chapter for further details.

The second system consists of two layers of woven materials, separated by a space of varying thickness (Figure 10.16b). The wall cavity between the layers of woven material is packed with straw-clay, clay-slip and wood chips, wool, or other materials to create an insulated wall structure. A cob-like mix (daub) is then applied to the woven lath surface. Plaster is then applied to the daub. Begin by making a wet mix, with characteristics of a discovery and infill coat — that is, adhesive but with a structural quality — and apply it so that it keys well into the daub, forming the base for subsequent layers of plaster. This plaster evens out and details the wall. Then comes a finish plaster. The finish coat can be made from earthen plaster, lime, or gypsum (interior only).

10-16 (a) (b): Wattle-and-daub consists of a matrix of bamboo or sticks to which mud is applied. (a) Single- or (b) double-walled structures are common. Double-walled systems can be filled with a variety of insulation materials, such as straw-clay or wool.

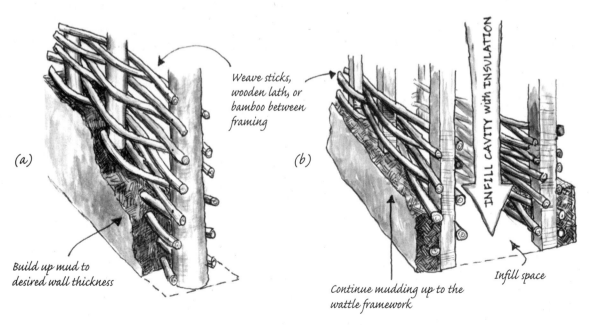

(a)

Build up mud to desired wall thickness

Weave sticks, wooden lath, or bamboo between framing

(b)

INFILL CAVITY WITH INSULATION

Continue mudding up to the wattle framework

Infill space

Plastering Rammed Earth Tire Walls

Earthships and other tire homes are built using automobile tires packed with dirt (figure 10-17). Earthships are typically built into hillsides (that is, they are earth-sheltered), so there's not much exterior plaster work on them.

The exterior plastering that is required — for example, on retaining walls — is often done with cement stucco. Because the exterior walls of most Earthships are unprotected and exposed to weather, earthen plasters are not generally recommended unless you're in a fairly dry climate or you can come up with a well-protected design, following guidelines presented in Chapter 3. Even then, you'll probably need to apply a protective alis and recoat the walls from time to time to maintain the surface. Perhaps the best option for weather-proofing exterior tire walls is to apply an earthen base coat, then a top coat of lime-sand plaster. Lime plaster is more resistant to direct weather, especially rain, and can provide a long-lasting and durable finish coat.

Because the interior walls are curved and rounded, earthen plasters produce exquisitely beautiful walls (see color gallery). In skilled hands, Earthship interior walls can be quite attractive. Interior tire walls are not exposed to weather and can be plastered with mud, cement, lime-sand, or even gypsum plaster.

Mud plaster and cement stucco are the most commonly used plaster products in modern Earthships. Earthen plasters produce a beautiful, soft and inviting finish with excellent acoustics. Cement stucco produces a colder, less visually inviting interior surface than other types of finishes. Rammed earth tire walls themselves very likely have a limited ability to breathe, and cement stucco cuts down on breathability even more. In addition, cement stucco wicks moisture into walls. These factors can result in a build up of moisture in tire walls which could cause problems over time. To prevent moisture from seeping into walls from the ground outside the structure, bermed exterior walls should be well drained and covered with plastic.

As noted earlier, earthen plaster can create a beautiful interior wall finish for Earthships and other rammed earth tire homes. Earthen plaster is better from an environmental and a health perspective, as noted in Chapters 2 and 4.

10-17: Worker packs tires with pneumatic tamping device.

WHAT IS AN EARTHSHIP?

The term Earthships refers to highly independent homes made from used automobile tires packed with dirt. Earthships are designed to capture and purify rainwater from the roof for domestic uses, recycle gray water as well as black water, and rely primarily on solar energy for heating and electricity. These naturally cooled, typically earth-sheltered homes are the invention of architect and builder Michael Reynolds of Taos, New Mexico. Not all rammed earth tire homes, however, are technically called *Earthships*.

Lime also makes a really good interior plaster for Earthships, because most of them contain a considerable number of plants. Plants are typically watered with gray water — that is, waste water from showers, sinks, and washing machines — which is fed into interior planters located along the south-facing glass. Extensive vegetation combined with earth sheltering result in moist interior environments: the higher the humidity, the more likely mold and mildew will form on the walls unless the home is well ventilated.

Earthen or gypsum plasters can be treated with natural linseed oil finishes, which resist mold. Limewashes applied to an earthen plaster also resist mold and mildew. Adding borax dissolved in warm water to the finish coat or to an alis applied over an earthen plaster finish has a similar effect.

Plastered tire walls tend to dry slowly because the walls don't breathe well. Be sure to ventilate well during plaster application. If you find sections where mold forms on the surface, spray with standard hydrogen peroxide and increase the circulation and ventilation throughout the area. Add heat if necessary. If moisture is high inside the home, mold may continue to be a problem after the walls have dried. Being aware of these problems, many people are now constructing rammed earth tire homes without the major sources of moisture or designing them to incorporate better ventilation systems to reduce interior moisture.

Preparing Tire Walls for Plaster

In a tire home, automobile tires are laid in a running bond, one row at a time. Each tire is packed with dirt — about 300 to 350 pounds of it per tire! As a wall rises, builders take special precautions to ensure that each course of packed tires is level and that the wall remains plumb as it increases in height. When finished, however, the walls are riddled with deep, V-shaped cavities located between adjacent tires, which must be packed with mud to smooth out the surface (Figure 10-18). Although the first Earthship design manual advocated the use of aluminum cans to help fill the voids, we don't recommend this, as aluminum is an extremely high-embodied-energy material.

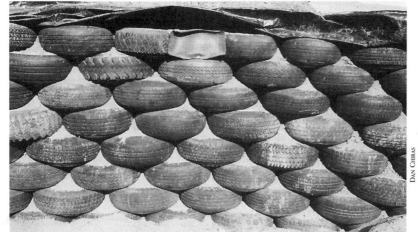

10-18: Walls of Earthships are highly indented. Mud needs to be packed in the V-shaped cavities between adjacent tires before plastering can begin.

DAN CHIRAS

(The electricity that goes into making an aluminum can from raw ore would run a 100-watt light bulb for four hours!) Aluminum is readily recyclable, too, so let it go back into production. Instead, you can use rocks from the site to help fill the voids. First mud is thrown into the void, then a rock or two can be jammed into the mud, followed by more mud.

Mud used to fill the voids typically consists of one part dirt from the site and one part coarse sand with water. Four large, double handfuls of straw are added per cement mixer load. Experiment with your soil: you want an adhesive but stable mix. If the dirt on site has a high sand content, you may not need to add any sand at all. Mix the sand, dirt, and water until the mix is fairly soupy (like a brownie batter). Add straw until it thickens, forming a firm packing material. The mud mix is typically tossed into the voids, a double handful at a time. If it contains enough clay, it will stick tenaciously. If it doesn't, or if the mix is too wet, the mud won't stay in place. Be sure to wet the tires before applying the mud.

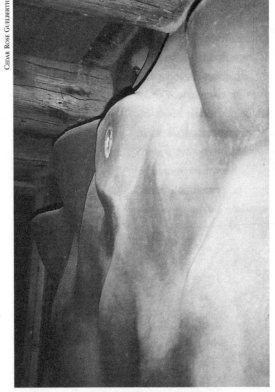

10-19: By not completely filling the voids between tires of an Earthship or tire home, you can produce a bumpy or wavy wall surface, as illustrated here.

Packing dirt in the voids takes place in two or three rounds. The first round (using mud and rocks) only partially fills the voids. After the first application of packing mud has dried completely, wet it down and apply a second layer of mud and rocks. Use mud of the same consistency as the mud used in round one. Let this material dry, then add more. However, for the final round, use a slightly sticky infill mix (described in Chapter 6) to even out and shape the surface. At this point, the mud should be flush with the surface of the tire treads (or what's left of them). When you're done, the wall should be fairly well covered with mud packing material, presenting a fairly uniform plane. Small sections of each tire tread will be exposed, but don't worry: they will be covered by the first coat of plaster.

Mud can be applied to create a bumpy or wavy surface that reveals the tires. This is achieved by not filling in the voids so thoroughly (Figure 10-19). It's up to you. Use an infill plaster mix to shape and soften the curves and to create a durable plaster surface.

Before applying the first coat of plaster — the scratch coat — cover any wood that will be plastered, and protect it with an adhesion coat. Paint on a thick layer of flour paste, manure, and coarse sand, as described in Chapter 6. You can also glue and staple burlap onto the wood to create a natural lath. An adhesion coat will also help hold burlap in place. Prior to packing, you can coat exposed tires with a flour-paste adhesion coat to enhance the bonding of the mud to the rubber. Or you can coat the exposed surface of the tires with a mixture of sand and AFM Safe Seal to seal the tires against potential release of chemical vapor and to create a rougher surface to key into the plaster. Although mud plaster generally adheres well to rubber, prep coats are especially helpful over bald tires, providing an extra level of security.

Plastering a Tire Wall

After the walls have been prepped, it is time to plaster. Apply three coats of plaster: a scratch coat, a brown coat, and a finish coat. Use the guidelines we've introduced in Chapter 6 as your starting point for mixes or contact a local Earthship builder, an experienced mudder who has plastered over tires, or an experienced local owner-builder for advice.

The scratch coat is applied over the exposed tire tread (and over the mud you've packed into the voids between tires), creating a continuous surface approximately ½-inch thick. However, don't begin work on the scratch coat until the packing mud has thoroughly dried; then wet the surface of the wall and begin plastering (Figure 10-20).

The scratch coat can be applied by hand or trowel. Hand application is quick and easy. Furthermore, it can result in a well keyed-in surface (with "tooth" or texture) to which the next coat adheres. Hand application also allows the mudder a higher degree of control — so he or she can press the plaster better into the tire wall.

Trowels can be used to apply the scratch only if the surface is even and consistent. For curves and dips, hand application is far superior. When using a trowel, be sure to scratch the surface of the plaster with horizontal grooves while the plaster is still a bit wet, as seen in Figure 8-10. The drier it gets, the harder it is to scratch. You may find it advantageous to scratch as you go — that is, scratch small sections so they don't dry out too much.

The next coat, the brown or infill coat, is applied over the scratch coat to achieve the desired wall surface. Be sure the scratch coat is completely dry before beginning work. Moisten then apply the brown coat by hand or by trowel, or a combination of the two. Moisten small sections, so the wall doesn't dry out before you get to it with your plaster; otherwise, you'll be wasting a lot of time misting and re-misting your walls.

10-20: Hand application of the packing material into the voids and hand application of the scratch coat (shown here) over tire walls is quick and easy, and produces a surface with "tooth" to which the next coat adheres well. Hand application also results in a higher degree of control, allowing mudders to key the plaster into the wall with greater success than a trowel allows.

CEDAR ROSE GUELBERTH

For tire walls, the infill (brown) coat is critical, as it determines whether the walls are curved, wavy, or flat. Remember, too, the surface of the infill coat should allow the finish coat to go on easily.

Next comes the finish coat. Begin by moistening the infill coat and then apply the finish coat by trowel or by hand — usually no more than ¼-inch thick. (Advice on applying a finish coat is given in Chapter 6.) Some mudders apply either a lime or a gypsum finish coat over an earthen brown (infill) coat. Earthen, lime, and gypsum finish coats can be pigmented to give them color or can be left *au naturel*. If you don't like the natural color of an earthen finish or want to brighten up a room, you can also apply an alis, natural paint, or a lime wash to it. Alises are described in Chapter 7 and lime washes, which are also easy to make and which provide color as well as excellent protection, are discussed in Chapter 8. Interior earthen finish plasters can be finished with natural oils, beeswax, or natural oil glazes with pigments.

Plastering Earthbag Walls

Earthbag construction is a relative newcomer to the natural building scene. Buildings are built from slightly moistened dirt packed into burlap or polypropylene sand or grain bags, as noted in Chapter 1. The polypropylene bags commonly used for earthbag construction are sand bags or those in which bulk grains are sold (Figure 10-21).

Earthbag walls are amenable to a variety of plasters. Two of the most commonly used are earthen and lime-sand plasters. Kaki Hunter and Doni Kiffmeyer, experienced earthbag builders and authors of *Earthbag Construction,* typically build in desert climates. For external plasters, they apply a base coat of clay-rich earthen plaster and a top coat of lime-sand plaster. Earthen, lime, and gypsum plasters are all suitable for the interior walls of earthbag structures.

Contrary to what you might think, plastering an earthbag wall is pretty simple. First, the surfaces of completed walls are fairly smooth, so you won't have to do a lot of infilling or shaping. Two to three coats of plaster are all that's required: you'll generally need an adhesive infill coat to smooth out the wall, and a finish coat. Second, there's very little wall preparation required

KAKI HUNTER

10-21: Carol Escott tamping earthbags in the construction of her home in the Bahamas. Earthbags are generally filled with moist, clay-containing soil, then tamped to create stable walls and a stable surface for plastering.

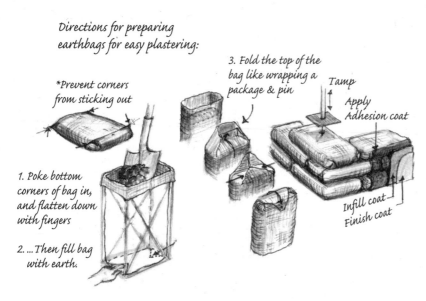

Directions for preparing earthbags for easy plastering:

*Prevent corners from sticking out

1. Poke bottom corners of bag in, and flatten down with fingers

2. ... Then fill bag with earth.

3. Fold the top of the bag like wrapping a package & pin

Tamp

Apply Adhesion coat

Infill coat
Finish coat

10-22 (a) (b): Diddling bags. (a) Pointed corners make plastering difficult; (b) clean, folded corners as shown here greatly facilitate plastering.

10-23: Earthbag wall ready for plastering. Note how well the corners of the bags are tucked in to create an even, easy-to-plaster surface.

CAROL ESCOTT

if you're applying an earthen plaster. Be sure to "diddle" the corners of your bags, says Hunter, or "you'll pay for it later with gobs more plaster." (For non-dirtbaggers, "diddling" is a term Hunter and Kiffmeyer use for folding in the bottom edges of each bag when they're being filled. This simple procedure creates "a crisp, solid edge" along the bottom of each bag, which eliminates soft spots in the wall and produces a firm, smooth surface for plastering (Figure 10-22). Check out their book for details. Third, clay-rich earthen plasters like those used for straw bale walls and other natural homes adhere well— extremely well — to burlap and polypropylene bags.

Earthen plasters can be applied by hand or trowel. Use a mix similar to the infill coat of a straw bale home for the first coat. To be sure the mix adheres well to the surface of the bags, you may need to increase the amount of clay. The mix needs to have structure for shaping (described in Chapter 6). You can modify the mix by adding various substances like flour paste, powdered milk, and cactus juice for durability, workability, and increased adhesion. Once the infill coat dries, wet the surface then apply the finish coat.

Lime-sand plasters don't adhere as well to earthbags as earthen plaster, but can be applied over an earthen base coat. To apply lime plaster alone, you'll need to attach stucco netting — one-inch chicken wire tightly secured to the dirt-filled bags. (Chicken wire can be attached with 2½-inch roofing nails driven into the tamped earthbags or attached to tie wires installed during wall construction between adjacent courses of bags. Tie wires are attached to the 4-point barbed wire laid down between courses of earthbags.) Once the stucco netting is in place, you simply trowel on the lime plaster.

APPLYING AN ADHESION COAT

You can apply an adhesion coat (Chapter 6) to an earthbag wall before coating it with an earthen plaster, but this is not generally required. Earthen plasters stick extremely well to earthbags as is.

Plastering Hybrid Buildings

There was a time when a straw bale house was a straw bale house and a cob house was a cob house, but those days are quickly vanishing as natural builders discover the benefits of combining natural material systems to achieve optimal performance. The result of their expanded horizons is that many natural homes are actually hybrids — that is, they contain two or more natural building systems. For instance, straw bale walls may be used to build the exterior envelope of a home if insulation is required. Adobe

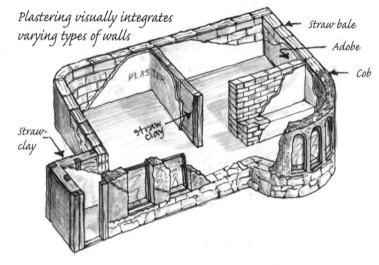

Plastering visually integrates varying types of walls

Straw-clay

PLASTER

straw clay

Straw bale
Adobe
Cob

10-24: Many natural builders are combining natural building systems to maximize building performance, as shown here. This introduces different wall surfaces that may require different approaches when it comes to plastering. Apply the details unique to the surface for wall preparation and base coats. Final finish coats tie the wall systems together.

or cob may be used to construct interior mass walls — ideal for passive solar heating. Cob benches might be installed and straw-clay may be used to make soundproof interior walls.

Hybrid homes present a bit of a challenge to the plasterer. Straw bale walls, for example, require a slip coat, then three coats of earthen plaster. Adobe and cob walls generally only require two coats of plaster, a base coat and a finish coat. Straw-clay walls would require the same. What's a person to do?

Treat each wall independently initially (Figure 10-25). That is, plaster each wall independently as we've outlined in the book *until* you get to the finish coat. Apply a slip coat and then the discovery and infill

CEDAR ROSE GUELBERTH

10-25: Here's an example of a hybrid structure showing the intersection of a straw bale wall coated with clay slip and a straw-clay wall (partially plastered).

coats of plaster to straw bale walls. Apply a scratch coat to the adobe, cob, and straw-clay walls. The finish coat, however, is the unifying layer of plaster: use the same finish to cover the entire structure. In other words, the finish coat is a common coat that runs over all of the walls. Alis, colored clay finishes, natural paints, or oil glazes can then be applied.

Quick Recap

Natural plasters work well on natural buildings, both straw bale and earthen walls, although the amount of preparation and plastering varies. The rougher or more irregular the wall surface, the more preparation and plastering will be required. In our view, cement and synthetic stucco should be avoided by natural builders. They can create many problems with moisture.

Study the material in this chapter, but be sure to read more and enroll in a workshop or two that teaches plastering techniques for the type of natural home you are interested in building.

In closing, natural plastering is a practical art that can be learned by anyone. Some people take to it like ducks to water: they are natural plasterers with an innate feel for the medium. Others need more training and experience to become comfortable and adept. If you are one of the former, congratulations. If you're not, be advised that with a little practice and study you, too, can become a gifted plasterer, carrying on a tradition that is easy on you and on Mother Earth.

H A P P Y   P L A S T E R I N G !

Natural Plaster Resource Guide

This resource guide begins with the most relevant resources — that is, those that cover the main topics discussed in this book. The second portion lists important natural building resources. Although they focus primarily on natural building materials and techniques, many of these resources also provide valuable information on natural plasters.

Plasters Publications and Organizations

Plastering (General References)

Chiras, Dan. "Perfect Plaster: A Homeowners Guide," *Mother Earth News*, Feb./Mar. 2003, pp. 64-71. Describes all plasters, but focuses on natural plasters.

Building Advisory Service and Information Network. *Mud Plasters and Renders.* This publication can be obtained from Taylor Publishing (listed below in Plaster and Natural Building Books — Publishers and Suppliers).

MacDonald, Alex C. *An Historical Digest of Plastering.* The California Plasterer, 1931 - 1937. A 28-article series on the history of plasters, available from Taylor Publishing.

Stagg, William D. and Brian F. Pegg. *Plastering: A Craftsman's Encyclopedia.* BSP Professional Books, 1989. Detailed coverage of conventional plastering.

Taylor, J.B. *Plastering.* Longman Scientific and Technical, 1990. Offers good advice on plastering technique.

Van Den Branden, F. and Thomas L. Hartsell. *Plastering Skills.* American Technical Publishers, 1984. General reference on conventional plastering.

Earthen Plasters and Clay Finishes, Including Alises

As you can see, there aren't many published resources on earthen plasters. Be sure to check out books on various natural building techniques and materials. Many of them contain chapters on plastering. Also, be sure to check out the workshops listed in this Resource Guide.

Chiras, Dan. "Get Plastered," *Natural Home*, May/June, 2002, pp 48 – 53. General description of natural plaster options, including the pros and cons of each one.

Crews, Carole. "Earth Plasters" in *Alternative Construction: Contemporary Natural Building Methods*, Lynne Elizabeth and Cassandra Adams, eds. New York: Wiley, 2000, pp. 255 - 263. Brief overview of earthen plasters.

Crews, Carole. "Earth Plasters and Aliz" in *The Art of Natural Building: Design, Construction, Technology*, Joseph F. Kennedy, ed. Kingston, NM: Networks Productions, 1999. A brief but useful article on alises by one of the world's experts on the subject.

Crews, Carole. "Alis, A Natural Polish; Clay finish for adobe structures," *Adobe Journal*, Issue 7, Autumn 1992, pp. 36 - 38.

Taylor, Charmaine. "Clay: The Eternal Stuff of Building. Clay for Plasters and Decorative Finishes," *Earthworks* Magazine, July/August 2000, pp. 19 - 22. Great resource.

Wanek, Catherine, ed. "Plasters," *The Last Straw: The International Journal of Straw Bale and Natural Building*, Issue 33, Spring, 2001. A complete issue of *The Last Straw* dedicated to the topic of plasters, mostly earth plasters.

Silicates, Natural Paints, and Finishes

Delamare, Francois and Bernard Guineau. *Colors: The Story of Dyes and Pigments.* New York, NY: Harry N. Abrams, 2000. Describes the history of natural pigments and dyes and other useful information.

Hermannsson, John. *Green Building Resource Guide*. Newtown, CT: Taunton Press, 1997. A gold mine of information on environmentally friendly building materials, including casein paints.

Homes, Dwight, Larry Strain, Alex Wilson, and Sandra Leibowitz. *GreenSpec: The Environmental Building News Product Directory and Guidelines Specifications*. Brattleboro, VT: E Build, Inc., 1999. Another gold mine of information on environmentally friendly building materials, including casein paints.

Innes, Jocasta. *Applied Artistry: A Complete Guide to Decorative Finishes for Your Home*. Boston: Little, Brown, and Company, 1995. A delightful, informative, and beautifully illustrated book on many different types of plasters.

Messmer, Reto, as transcribed by Carole Crews. "Natural Paints with Reto Messmer," *The Art of Natural Building: Design, Construction, Technology*, Joseph F. Kennedy, ed. Kingston, NM: Networks Productions, 1999. Notes from Reto's presentation at The Natural Building Colloquium in Kingston, New Mexico.

Sloan, Annie. *Paint Alchemy*. London: Collings and Brown, 2001. Very useful resource for those who want to learn more about traditional natural paints.

Sloan, Annie and Kate Gwyn. *Classic Paints and Faux Finishes*. London: Collins and Brown, 2001. Great reference on natural paints.

Lime Plaster and Lime Washes

Brenna, Rory. "Working with Lime Plaster," *Traditional Building*, November/December, 1998. Describes application of lime to Mt. Vernon, George Washington's home.

Devon Earth Building Association. *Appropriate Plasters, Renders and Finishes for Cob and Random Stone Walls in Devon*. Devon Earth Building Association, Devon, England, 1993. A brief but fact-filled guide on lime plaster for cob and stone walls.

Holmes, Stafford and Michael Wingate. *Building with Lime: A Practical Introduction*. London: Intermediate Technology Publications, 1997. A more advanced treatise on the use of lime, including lime plaster.

Hunter, Kaki and Doni Kiffmeyer. "The Discovery and Use of Carbide Lime," *The Last Straw: The Grassroots Journal of Straw Bale and Natural Building*, Issue 33, 2001, pages 17. Great overview of carbide lime and its use in lime-sand plaster.

Jones, Barbara. "Working with Lime in England," *The Last Straw: The Grassroots Journal of Straw Bale and Natural Building*, Issue 26, 1999, pages 22 - 25. Excellent resource on lime plasters.

Maxwell, Ingval. *Lime Technology Workshop on Rendered Facades*. Edinburgh, Scotland: Historic Scotland, 1998. Detailed information on maintaining and refinishing lime plasters.

Schofield, Jane. *Lime in Building: A Practical Guide*. Crediton, Devon, England: Black Dog Press, 1994. Good resource for those interested in learning more about lime plastering.

Schofield, Jane. "An Introduction to Lime," *Structural Survey* 13(2), 1995, pp. 4 - 5. A brief but informative survey of lime and its benefits.

Steen, Bill and Athena Steen, eds. "Lime Plaster," *The Last Straw: The Grassroots Journal of Straw Bale and Natural Building*, Issue 29, Spring, 2000. This entire issue of *The Last Straw* is dedicated to lime plasters and is a must-read for anyone interested in learning about lime plaster.

Taylor, Charmaine. *All About Lime: A Basic Information Guide*. Eureka, CA: Taylor Publishing, 1999. A fairly detailed and comprehensive booklet with lots of good information.

Ward, John D. and Ingval Maxwell, eds. *International Lime Conference Proceedings*. Edinburgh, Scotland: Building Limes Forum and Historic Scotland, 1999. A collection of 15 articles by various authors.

ORGANIZATIONS

National Lime Association, 200 Glebe Road, Suite 800, Arlington, VA 22203. Tel: (703) 243-5463. Sells books on lime plastering.

National Park Association. P.O. Box 25287, Denver, CO 80225. Tel: (303) 969-5440. Dozens of technical articles on building restoration including the use of lime plasters.

Gypsum Plaster

Books

Dorn, Gary. "Gypsum Plaster on Straw Bales," *The Last Straw: The Grassroots Journal of Straw Bale and Natural Building*, Issue 33, page 28.

Organizations

Gypsum Association, 810 Pine Street, NE, Suite 510, Washington, DC 20002. Tel: (202) 289-3707. <www.national-gypsum.com>. Offers information on gypsum plaster.

Publishers and Suppliers — Plaster and Natural Building Books

The Black Range Lodge/Natural Building Resources. 119 Main St., Kingston, NM 88042. Tel: (505) 895-5652. Videos and books on natural building.

Building for Health Materials Center. Offers numerous books on natural building and natural plaster. See listing under "Suppliers of Natural Plasters, Alises, Natural Paints, and Tools".

Chelsea Green. P.O. Box 428, White River Junction, VT 05001. Tel: (800) 639-4099. website <www.chelseagreen.com>. Publishes a wide assortment of books on natural building.

DAWN/Out on Bale by Mail. 6570 W. Illinois St., Tucson, AZ 85735. Sells numerous books on natural building.

Earthwood Building School, 336 Murtagh Hill Road, West Chazy, NY 12992. Tel: (518) 493-7744. Sells a variety of books on natural building, especially cordwood.

New Society Publishers. P.O. Box 189, Gabriola Island, BC VOR 1XO Canada. Tel: (250) 247-9737. website <www.newsociety.com>. Publishes books on straw bale construction, natural building, healthy building, and, of course, this book on natural plasters.

Solar Survival Architecture. P.O. Box 1041, Taos, NM 87571. Tel: (505) 751-0462. website <www.earthship.org>. Sells books on Earthship construction.

Taylor Publishing, P.O. Box 6985, Eureka, CA 95502. Tel: (707) 441-1632. website <www.dirtcheapbuilder.com>. Sells numerous of titles on natural building, including numerous books, pamphlets and articles on lime plaster.

Suppliers of Natural Plasters, Alises, Natural Paints, and Tools

Building for Health Materials Center. P.O. Box 113, Carbondale, CO 81623. Tel: (800) 292-4838 or (970) 963-0437. E-mail <Contactus@buildingforhealth.com>, website <www.buildingforhealth.com>. Offers a complete line of healthy, environmentally safe building materials. Specializes in straw bale construction products and natural plastering products including bulk clays, mica, pigments, lime putty; natural paints, oils, stains, and finishes; sealants; and tools, including trowels, stucco pumps, and texture guns. Offers special pricing for owner-builders and contractors.

Eco-Wise. 110 W. Elizabeth, Austin, TX 78704. Tel: (512) 326-4474. Retail store that carries casein paints and natural oils.

Environmental Building Supplies. 1331 NW Kearney Street, Portland, OR 97209. Tel: (503) 222-3881. Green building materials outlet for the Pacific Northwest. Carries a wide assortment of products including casein paints.

Environmental Construction Outfitters. 44 Crosby Street, New York, NY 10012. Tel: (800) 238-5008. Sells an assortment of green building materials, including casein paints.

Environmental Home Center. 44 Crosby Street, New York, NY 10012. Tel: (800) 238-5008. Offers a variety of green building materials, including casein paints.

Planetary Solutions. 2030 17th Street, Boulder, CO 80302. Tel: (303) 442-6228. Offers casein paints, natural oils, and an assortment of green building materials.

Mike Wye and Associates. Tel: 0149 281 644 website <www.mikewye.com>. Traditional and ecological building products in the United Kingdom, including lime putty and lime plaster.

Ochres and Oxides, Olde World Colors. 606 Commercial Avenue, Suite B, Anacortes, WA 98221. Tel: (360) 708-8992. website <www.ochresandoxides.com>. Owned and operated by Charlotte Underwood who distributes natural pigments (sixteen ochers and oxides) imported from Europe.

Workshops on Natural Plasters and Natural Building

Below is a list of major workshops in natural and plaster and natural building. For a listing of when workshops are offered, consult the most recent issue of The Last Straw *journal.*

Natural Plasters Workshops

Carole Crews. HC78 Box 9811, Rancho de Taos, New Mexico 87557. Tel: (505) 758-7251. E-mail <seacrews@taosnet.com>. Offers workshops on earthen plasters, alis and natural decorative finishes, and sculpting.

CRG Design — Designs For Living. P.O. Box 113, Carbondale, CO 81623 Tel: (970) 963-0437. Workshops on earthen plasters, natural clay finishes, and natural paints led by Cedar Rose Guelberth.

North American School of Natural Building. P.O. Box 123, Cottage Grove, OR 97424 Tel: (541) 942-3021 or (541) 942-2005. Ianto Evans and Linda Smiley offer workshops on natural building and earth plastering.

OKOKOK Productions. 256 E. 110 S., Moab, UT 84532. Tel: (435) 259-8378. Workshops on earthbag earthen and lime plasters. Doni Kiffmeyer and Kaki Hunter.

Bill and Athena Steen, The Canelo Project. HC 1, Box 324, Elgin, AZ 85611 Phone: (520) 455-5548. Earthen plaster and lime plaster workshops.

Solar Energy International. P.O. Box 715, Carbondale, CO 81623 Tel: (970) 963-8855. Offers a complete natural plaster workshop lead by Cedar Rose Guelberth.

Natural Building (General) Workshops

Build Here Now. P.O. Box 240, San Cristobal, NM 87564. Tel: (505) 586-1269. E-mail <info@lamafoundation.org>. website <www.lamafoundation.org>. This week-long annual affair at the Lama Foundation in northern New Mexico offers a variety of workshops on natural building with several workshops on earthen plaster, alis, and lime plaster.

CRG Design — Designs For Living offers natural building and design workshops led by Cedar Rose Guelberth. See contact information in previous listing.

DAWN/Out On Bale By Mail. 6570 W. Illinois St. Tucson, AZ 85735. Tel: (520) 624-1673 E-mail <dawnaz@earthlink.net>. website <www.greenbuilder.com/dawn/>. Books and workshops on natural building.

Emerald Earth Workshops. P.O. Box 764, Boonville, CA 954`5. Tel: (707) 895-3302. E-mail <lorax@ap.net>. Michael Smith offers workshops on natural building and permaculture. P.O. Box 764, Boonville, CA 954`5. Tel: (707) 895-3302.

Fox Maple School of Traditional Building. P.O. Box 249, Brownfield, ME 04010 Tel: (800) 369-4005 or (207) 935-3720. Frank Andersen offers workshops on cob and thatching.

Real Goods Institute for Solar Living. P.O. Box 836, Hopland, CA 95449 Tel: (800) 762-7325. Offers a variety of workshops on natural building techniques, including natural plaster and some general workshops that cover several topics.

Solar Energy International. P.O. Box 715, Carbondale, CO 81623 Tel: (970) 963-8855. Their natural house building and solar design workshop covers several natural building techniques.

Shay Solomon, 1050 S. Verdugo, Tucson, AZ 85745. Workshops on natural earth building and straw bale construction.

SunEarth Construction. Keith and Tracy Lindauer, P.O. Box 113, Rico, CO 81332 Tel: (970) 967-2882. Periodically has offered a general workshop on natural building techniques and other important topics.

Sustainable Systems Design. 9124 Armadillo Trail, Evergreen, CO 80439. Tel: (303) 674-9688. E-mail <danchiras@msn.com>. Workshops on natural plasters and finishes and straw clay construction led by Dan Chiras.

Straw Bale Workshops

The Canelo Project, HC 1, Box 324, Elgin, AZ 85611 Phone: (520) 455-5548. Bill and Athena Steen offer straw bale and plaster workshops.

Chris Magwood, Camel's Back Construction. 2648 Cooper Road, RR 3, Madoc, Ontario K0K 2K0 Canada. Tel: (613) 473-1718. Workshops on straw bale construction and much more.

Catherine Wanek and Pete Fust, Black Range Lodge, Star Route 2, Box 119, Kingston, NM 88042, Tel: (505) 895-5652. Hosts workshops on natural building.

Carol Escott and Steve Kemble, Sustainable Support Systems, P.O. Box 318, Bisbee, AZ 85603. Tel: (520) 432-4292. Workshops on straw bale construction.

Center for Alternative Technology. Machynlleth, Powys SY20 9AZ Tel: 01654 703409. This educational group in the United Kingdom offers workshops on straw bale construction and many other subjects.

Real Goods Institute for Solar Living. P.O. Box 836, Hopland, CA 95449 Tel: (800) 762-7325. Offers a variety of workshops on straw bale construction, timber frame, and natural plaster.

Yellow Mountain Institute for Sustainable Living. P.O. Box 205, Batesville, Virginia 22924 Tel: (540) 456-6447 or (804) 963-6107. On-line workshop on straw bale home construction.

Rammed Earth Building Workshops

Rammed Earth Works. 101 S. Coombs, Suite N, Napa, CA 94559 Tel: (707) 224-2532. Workshops and a host of other services. 101 S. Coombs, Suite N, Napa, CA 94559 Tel: (707) 224-2532

Real Goods Institute for Solar Living. P.O. Box 836, Hopland, CA 95449 Tel: (800) 762-7325. Rammed earth building workshops and workshops on a host of other subjects.

Earthships and Rammed Earth Tire Homes Workshops

Earthship Biotecture. P.O. Box 1041, Taos, NM Tel: (505) 751-0462. Offers training in Earthship construction and a host of other services as well.

Yellow Mountain Institute for Sustainable Living. P.O. Box 205, Batesville, Virginia 22924 Tel: (540) 456-6447 or (804) 963-6107. On-line workshop on rammed earth tire construction.

Adobe Workshops

The Earth Building Foundation. 5928 Guadalupe Trail NW, Albuquerque, NM 87107. Tel: (505) 345-2613. Offers several workshops on adobe construction.

Cob Workshops

The Cob Cottage Company. P.O. Box 123, Cottage Grove, OR 97424 Tel: (541) 942-3021 or (541) 942-2005. Offers workshops on beginning and advanced cob construction and other subjects.

Groundworks. P.O. Box 381, Murphy, OR 97533. Offers workshops on cob in different locations.

Straw/Clay Workshops

Robert Laporte. P.O. Box 864, Tesuque, NM 87574. Tel: (505) 989-1813. Workshops on straw clay and timber frame construction.

Earthbag Building Workshops

Cal Earth. 10376 Shangri-La Avenue, Hesperia, CA 92345 Tel: (760) 244-2201. Offers a week-long apprenticeship retreat in earthbag (superadobe) construction.

OKOKOK Productions. 265 E. 110 S., Moab, UT 84532. Tel: (435) 259-8378. Workshops on earthbag construction and plastering by Doni Kiffmeyer and Kaki Hunter.

Additional Resources

This section contains important resources on natural building, many of which provide information on natural plasters.

Natural Building (General References)

Chiras, Daniel D. *The Natural House: A Complete Guide to Healthy, Energy-efficient, Environmental Homes*. White River Junction, VT: Chelsea Green, 2000. The best-selling guide to natural building by the co-author of this book. Contains a wealth of information on natural building and a good overview of plastering.

Elizabeth, Lynne and Cassandra Adams. *Alternative Construction: Contemporary Natural Building Methods*. New York: Wiley, 2000. A compilation of articles on numerous natural building techniques with a little information on natural plasters.

Kennedy, Joseph F., Michael G. Smith, and Catherine Wanek, eds. *The Art of Natural Building: Design, Construction, and Technology*. Gabriola Island, BC: New Society Publishers, 2002. Contains an assortment of articles on natural building, including natural plasters and alises, from experts in the field.

Minke, Gernot. *Earth Construction Handbook: The Building Material Earth in Modern Architecture*. Southampton, England: WIT Press, 2000. Contains a great deal of technical information and a good section of earth plasters, called loam plasters.

MAGAZINES

Mother Earth News, 1503 SW 42nd Street, Topeka, KS 66609. Tel: (785) 274-4300. website <www.motherearthnews.com>. Publishes articles on natural building, natural plasters, renewable energy, and other important topics.

Natural Home, 201 E. 4th Street, Loveland, CO 80537. Tel: (800) 340-5486. website <www.naturalhomemagazine.com>. Covers a wide range of topics vital to healthy, natural building.

Straw Bale Construction

BOOKS

Eisenberg, David. *Straw Bale Building and the Codes: Working with Your Code Officials*. Tucson, AZ: DCAT, 1996. Valuable resource for those whose building departments have never approved or heard of straw bale construction. To order a copy: <www.strawbalecentral.com>.

Haggard, Ken and Scott Clark, eds. *Straw Bale Construction Details: A Sourcebook*. Angels Camp, CA: California Straw Bale Association, 2000. Available from CASBA, P.O. Box 1293, Angels Camp, CA 95222. Tel: (209) 785-7077. Contains many important details on straw bale construction but no information on plastering.

Kemble, Steve and Carol Escott. *How to Build Your Elegant Home with Straw* (manual and video set). Bisbee, AZ: Sustainable Systems Support, 1995. To order contact authors at: Sustainable Systems Support, P.O. Box 318, Bisbee, AZ 85603.

King, Bruce. *Buildings of Earth and Straw*. Sausalito, CA: Ecological Design Press, 1996. A great book for the technically-minded reader. Contains a wealth of information on tests run on straw bale structures.

Lacinski, Paul and Michel Bergeron. *Serious Straw Bale: A Home Construction Guide for All Climates*. White River Junction, VT: Chelsea Green, [2001]. Contains a great deal of information on building with straw bales and plastering in cold and wet climates. Detailed coverage of lime plaster.

Magwood, Chris and Peter Mack. *Straw Bale Building: How to Plan, Design, and Build Straw Bale*. Gabriola Island, British Columbia, Canada: New Society Press, 2000. A wonderfully written book on building straw bale in a variety of climates, especially northern climates. Contains a fair amount of information on plastering.

Myhrman, Matts and S.O. Myhrman. *Build It with Bales (Version 2.0): A Step-by-Step Guide to Straw-bale Construction*. Tucson: Out on Bale, 1998. A superbly illustrated and recently updated manual on straw bale construction. Contains a fair amount of information on wall preparation, plasters, and plastering.

Steen, Athena S., Bill Steen, David Bainbridge, and David Eisenberg. *The Straw Bale House*. White River Jct., VT: Chelsea Green, 1994. The best-selling book that helped fuel interest in straw bale construction. Contains a wealth of information on straw bale construction, wall preparation, and plasters.

Straw Bale Testing documents. Hard copy reports on code testing of straw bale walls. May be essential to help your building department understand what you are doing. A must-read for architects and builders and building department officials. To order a copy of all three reports: <www.strawbalecentral.com>.

MAGAZINES AND NEWSLETTERS

The Last Straw. Joyce Coppinger, 2110 South 33rd St., Lincoln, NE 68506. Tel: (402) 483-5135. website <www.strawhomes.com>. Quarterly journal containing the latest information on straw bale construction. Annual resource issue contains a gold mine of information. Publishes articles on natural plasters. This is an absolute must for all straw bale enthusiasts!

VIDEOS

Building with Straw, Vol. 1: A Straw-Bale Workshop. Black Range Films, 1994. Documents a weekend workshop in which volunteers helped to build a two-story greenhouse addition onto a lodge. To order: <www.strawbalecentral.com>.

Building with Straw, Vol. 2: A Straw-Bale Home Tour. Black Range Films, 1994. A tour of ten straw bale structures in New Mexico and Arizona. To order: <www.strawbalecentral.com>.

Building with Straw, Vol. 3: Straw-Bale Code Testing. Black Range Films, 1994. Takes you on a tour of ten straw bale structures in New Mexico and Arizona. Presents the insights of the owners/builders. To order: <www.strawbalecentral.com>.

How to Build Your Elegant Home with Straw Bales. Covers the specific of building a load-bearing straw bale home. Comes with a manual. To order: <www.strawbalecentral.com>.

The Straw Bale Solution Narrated by Bill and Athena Steen and produced by Catherine Wanek. Features interviews with architects, engineers, owner-builders. Covers basics of straw bale construction and much more. You can order directly from the Steens by contacting their website <www.caneloproject.com> or by contacting the producer at: <www.strawbalecentral.com>.

ORGANIZATIONS

Austrian Straw Bale Network, A-3720 Baierdorf 6, Austria. website <www.baubiologie.at>.

California Straw Building Association. 115 Angelita Avenue, Pacifica, CA 94044. Tel: (805) 546-4274. website <www.strawbuilding.org>. This group is involved in testing straw bale structures. They also offer workshops and sponsor conferences.

The Canelo Project. HC 1, Box 324, Elgin, AZ 85611. their website <www.caneloproject.com>. Founded and run by Athena and Bill Steen, co-authors of *The Straw Bale House*. They offer workshops, videos, and books on straw bale construction as well as information on building codes and results of tests on straw bale homes.

Center for Maximum Building Systems. 8604 FM 969, Austin, TX 78724. Tel: (512) 928-4786. Working at the cutting edge of building materials, systems, and methods. Led by Pliny Fisk, III.

Development Center for Appropriate Technology. P.O. Box 27513, Tucson, AZ 85726-7513. Tel: (520) 624-6628. website <www.dcat.net>. Offers a variety of services including consulting, research, testing, assistance with code issues, project support, instruction, and workshops. Contact them in care of David Eisenberg.

European Straw Bale Building Network. Contact them at <strawbale-1@eyfa.org>.

Greenfire Institute, Ted Butchart, 1509 Queen Anne Ave. N #606, Seattle, WA 98103. Offers straw bale workshops, design consultation, full design, building consultation, and full building options, all using straw or other sustainable materials. E-mail <greenfire@delphi.com>.

Japan Straw Bale House Association, 8-9 Honcho, Utsunomiya, Tochigi, Japan 3200033. website <www.geocities.co.jp/NatureLand/1946/>.

MidAmerica Straw Bale Association. 2110 South 33rd Street, Lincoln, NE 68506. website <www.strawhomes.com/sban>. E-mail <jc10508@alltell.net>. Publishes a quarterly newsletter and calendar of workshops. Offers hands-on-training, tours, and much more.

Natural Building Resource Center. Rt. 1, Box 245B, Mauk, GA. website <www.gnat.net/~goshawk>. Assists and supports owner-builders interested in straw bale, cob, and earthbags.

Norwegian Straw Bale Building Organization, Waemhus, N-1540 Vestby, Norway. E-mail <arild.berg3@chello.no>.

Straw Bale Association of Nebraska. 2110 S. 33rd St., Lincoln, NE 68506-6001. Tel: (805) 483-5135. Active in promoting straw bale construction.

Straw Bale Association of Texas. P.O. Box 4211, Austin, TX 78763. Tel: (512) 302-6766. website <www.greenbuilder.com/sbat/>. Sponsors monthly meetings, publishes a newsletter, and provides a host of other resources.

Straw Bale Building Association for Wales, Ireland, Scotland, and England. SBBA, P.O. Box 17, Todmorden, OLI1 8fD, England. Tel: 00-44-1706-818126. Exchanges information and experience in straw bale construction.

Straw Bale Building Association of Australia. Contact at <sbaoa@yahoo.com.au>.

Straw Bale Construction Association of New Mexico. Contact Catherine Wanek, Route 2, Box 119, Kingston, NM 88042. Tel: (505) 895-5652. E-mail <blackrange@zianet.com>

Cob Building

BOOKS

Bee, Becky. *The Cob Builders Handbook: You Can Hand-Sculpt Your Own Home.* Murphy, OR: Groundworks, 1997. Amply illustrated and clearly written introduction to cob building with a brief section on plasters and plastering.

Denzer, Kiko. *Build Your Own Earth Oven*. Blodgett, OR: Hand Print Press, 2001. Describes how to build an oven out of cob. Order from Kiko at P.O. Box 576, Blodgett, OR 97326. Tel: (541) 438-4300.

Evans, Ianto, Michael G. Smith, and Linda Smiley. *The Hand-Sculpted House: A Practical and Philosophical Guide to Building a Cob Cottage*. White River Junction, VT: Chelsea Green, 2002. Superb resource! A must-read for anyone interested in cob building.

Smith, Michael G. *The Cobber's Companion: How to Build Your Own Earthen Home*. 2nd Edition. Cottage Grove, OR: The Cob Cottage Company, 1998. Well written introduction to cob; many excellent and useful illustrations.

MAGAZINES AND NEWSLETTERS

The CobWeb. The only cob-focused periodical. Published twice yearly by The Cob Cottage Company. P.O. Box 123, Cottage Grove, OR 97424. website <www.deatech.com/cobcottage/>

VIDEOS

Building with the Earth: Oregon's Cob Cottage Co. Great resource. Obtain from The Cob Cottage Company.

ORGANIZATIONS

Center for Alternative Technology. Machynlleth, Powys SY20 9AZ. Tel and fax: 01654 703409. Check out their website for a list of offerings at <www.cat.org.uk/>. This educational group in the United Kingdom offers workshops on earth building and natural finishes, among other topics.

Adobe

BOOKS

Bourgeois, Jean-Louis. *Spectacular Vernacular: The Adobe Tradition*. New York: Aperture Foundation, 1989. Superb and beautifully photographed overview of adobe building throughout the world.

McHenry, Jr., Paul G. *Adobe and Rammed Earth Buildings: Design and Construction*. Tucson, AZ: University of Arizona Press, 1984. Excellent reference, covering history, soil selection, adobe brick manufacturing, adobe wall construction, and many more topics. Good coverage of earthen plastering.

McHenry, Jr., Paul G. *Adobe: Build it Yourself*. Tucson, AZ: University of Arizona Press, 1985. Highly readable and surprisingly thorough introduction to many aspects of adobe construction. Focuses on cement and gypsum plaster.

Stedman, Myrtle and Wilfred Stedman. *Adobe Architecture*. Santa Fe, NM: Sunstone Press, 1987. Contains numerous drawings of houses, floor plans, and well-illustrated basic information on making adobe bricks and laying up walls.

ORGANIZATIONS

The Earth Building Foundation, Inc. website http//<www.earthbuilding.com>.

Formerly the Earth Architecture Center, International, Ltd. A nonprofit organization whose mission is to help people learn how to utilize earth building, especially adobe and rammed earth. Offers a newsletter, publications, information on building codes, workshops and training. Especially helpful is an extensive search list of approximately 1,300 references.

MAGAZINES

Inter Americas Adobe Builder Magazine. P.O. Box 153, Bosque, NM 87006. Tel: (505) 861-1255 website: <www.adobebuilder.com>. This magazine, while focusing primarily on adobe, offers articles on rammed earth from time to time

Rammed Earth

BOOKS

Berglund, Magnus. *Stone, Log, and Earth Houses: Building with Elemental Materials*. Newtown, CT: Taunton Press, 1986. Chapters 9 - 12 of this book provide technical information on rammed earth construction and some beautiful photos.

Easton, David. *The Rammed Earth House*. White River Junction, VT: Chelsea Green, 1996. An informative, highly readable book. A must for anyone considering this building technique. No discussion of plaster or plastering.

King, Bruce. *Buildings of Earth and Straw*. Sausalito, CA: Ecological Design Press, 1996. Another essential reading for anyone interested in building a rammed earth home.

Middleton, G. F. *Earth Wall Construction*. Bulletin #5. North Ryde, NSW, Australia: CSIRO-DBCE, 1995. A manual on rammed earth showing a unique forming system. Appendices contain structural and insulation calculations.

VIDEOS

Rammed Earth Construction. A 29-minute video produced by Hans-Ernst Weitzel. To order, call Bullfrog films at: (800) 543-3764.

The Renaissance of Rammed Earth. This 31-minute video features David Easton and serves as an excellent introduction to the subject or a companion to The Rammed Earth House. Available from Chelsea Green.

ORGANIZATIONS

The Earth Building Foundation, Inc. See description listed in adobe section.

Earthships and Tire Homes

BOOKS

Reynolds, Michael. *Earthship: Build Your Own. Vol. I.* Taos, NM: Solar Survival Press, 1990. A must read for those wanting to understand the basics of early Earthship design. This book contains some outdated information, however, so be sure to read the more current volumes and check out the *Earthship Chronicles* for up-to-date information.

Reynolds, Michael. *Earthship. Systems and Components. Vol. II.* Taos, NM: Solar Survival Press, 1990. Explains the various systems such as graywater, solar electric, and domestic hot water. Essential reading for all people interested in sustainable housing.

Reynolds, Michael. *Earthship: Evolution Beyond Economics. Vol. III.* Taos, NM: Solar Survival Press, 1993. Presents many of the new developments. Latest information, however, may be best learned in workshops, tours of new houses, and the *Earthship Chronicles*.

MAGAZINES AND NEWSLETTERS

Earthship Chronicles published by Earthship Global Operations, P.O. Box 2009, El Prado, NM 87529. Tel: (505) 751-0462. Pamphlets issued periodically to disseminate new information. You will find pamphlets on greywater, catchwater, blackwater, mass vs. insulation, and equipment catalog.

Solar Survival Newsletter. Available from Solar Survival Architecture, P.O. Box 1041, Taos, New Mexico 87571. E-mail: <solarsurvival@earthship.org>.

VIDEOS

Building for the Future. This is a video about the building of my house. It explains the how it was built and many green building products. Contact me at (303) 674-9688 or via E-mail <danchiras@msn.com>.

Dennis Weaver's Earthship. Shows construction of actor Dennis Weaver's Earthship. Well done and very informative. Helpful in securing building permits. Available from Solar Survival Architecture at their on-line store at <www.dennisweaver.com/habitat.html>.

The Earthship Documentary. Describes the history of Earthship construction, the underlying philosophy behind this unique structure, and building techniques. Available from Solar Survival Architecture at their on-line store (listed above).

Earthship Next Generation. A look at new Earthship designs and constructions. Available from Solar Survival Architecture at their on-line store (listed above).

From the Ground Up. Takes you through the process of building an Earthship. Available from Solar Survival Architecture at their on-line store (listed above).

Earthbags

BOOKS

Hunter, Kaki and Doni Kiffmeyer. *Earthbag Building*, (Gabriola Island, BC: New Society, 2003). Detailed book on earthbag construction with information on plastering. Informative, and well organized. Can be obtained from the authors at their company OKOKOK Production at 256 E. 100 South, Moab, UT 84532. Tel: (435) 259-8378. E-mail: <okokok@lasal.net>. website: <www.ok-ok-ok.com>

Wojciechowska, Paulina. *Building with Earth: A Guide to Flexible-Form Earthbag Construction*. White River Junction, VT: Chelsea Green, 2001. Describes earthbag construction and offers some details on plastering.

ORGANIZATIONS

CalEarth, California Institute of Earth Art and Architecture. Nader Khalili, CalEarth/Geltaftan Foundation, 10376 Shangri La Avenue, Hesperia, CA 92345. Tel: (760) 244-0614. website <www.calearth.org>. Offers information on earthbag construction, including an online newsletter.

Straw-Clay

BOOKS

Laporte, Robert. *MoosePrints: Holistic Home Building*. Santa Fe, NM: Natural House Building Center, 1993. A brief booklet on straw-clay construction, with some excellent illustrations.

Stone Building: Foundations

BOOKS

Cramb, Ian. *The Complete Guide to the Art of the Stonemason*. Cincinnati, OH: Betterway Publications, 1992. A detailed reference, ideally suited for the beginner.

Long, Charles. *The Stone Builder's Primer: A Step-by-Step Guide for Owner-Builders*. Willowdale, Ontario, Canada: Firefly Books, 1998. Contains the most complete instructions on stone home building of the books I've read or reviewed.

McRaven, Charles. *Stonework: Techniques and Projects*. Pownal VT: Storey Books, 1997. Covers the use of stone to build walls, paths, ponds, steps, and much more.

Vivian, John. *Building Stone Walls*. Pownal, VT: Storey Books, 1986. A well-illustrated and well-written book that focuses primarily on stone wall construction.

Index

About the Authors

Cedar Rose Guelberth has been a natural builder for over 25 years. She travels nationwide lecturing on natural plasters, natural building systems, healthy homes, and indoor air quality issues. She spent 3 years in Europe studying traditional and historic architecture and construction, with an emphasis on natural paints and plasters. Cedar Rose spends much of her time designing natural homes, teaching groups the art of natural plastering, and consulting with homeowners and contractors about healthy, environmental and natural construction. Her passion for natural building techniques has led her to author many articles about the subject. Currently she resides in Colorado.

Cedar Rose owns Building for Health Materials Center, a national retail company that carries an extensive supply of environmetal products including plastering materials.

Dan Chiras lives in a passive solar/solar electric home in the mountains of Colorado built from rammed earth tires and straw bales that is finished in part with earthen plasters. Dan is author of numerous books including the *The Natural House: A Complete Guide to Healthy, Energy-Efficient, Environmental Homes* and *The Solar House: Passive Heating and Cooling*. He is a contributing editor to *Mother Earth News*, and has published numerous other articles on environmental building and related subjects in *Mother Earth News*, *Natural Home*, and *The Last Straw*. Dan teaches courses on ecological design at Colorado College and lectures widely on sustainable building practices. He also teaches workshops on natural building and natural plasters, as well as passive solar heating and cooling.

If you have enjoyed *The Natural Plaster Book*, you might also enjoy other

BOOKS TO BUILD A NEW SOCIETY

Our books provide positive solutions for people who want to make a difference. We specialize in:

**Sustainable Living • Ecological Design and Planning • Natural Building & Appropriate Technology
New Forestry • Environment and Justice • Conscientious Commerce • Progressive Leadership
Educational and Parenting Resources • Resistance and Community • Nonviolence**

For a full list of NSP's titles, please call 1-800-567-6772 or check out our web site at:
www.newsociety.com

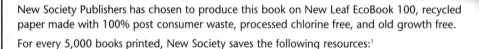

New Society Publishers

ENVIRONMENTAL BENEFITS STATEMENT

New Society Publishers has chosen to produce this book on New Leaf EcoBook 100, recycled paper made with 100% post consumer waste, processed chlorine free, and old growth free.

For every 5,000 books printed, New Society saves the following resources:[1]

22	Trees
2,006	Pounds of Solid Waste
2,207	Gallons of Water
2,878	Kilowatt Hours of Electricity
3,646	Pounds of Greenhouse Gases
16	Pounds of HAPs, VOCs, and AOX Combined
6	Cubic Yards of Landfill Space

[1]Environmental benefits are calculated based on research done by the Environmental Defense Fund and other members of the Paper Task Force who study the environmental impacts of the paper industry.

For more information on this environmental benefits statement, or to inquire about environmentally friendly papers, please contact New Leaf Paper – info@newleafpaper.com Tel: 888 • 989 • 5323.

NEW SOCIETY PUBLISHERS